ALTERNATIVE ENERGY:
SOURCES AND SYSTEMS

Donald Steeby

DELMAR
CENGAGE Learning

Australia • Brazil • Japan • Korea • Mexico • Singapore • Spain • United Kingdom • United States

**Alternative Energy: Sources
and Systems
Donald Steeby**

Vice President, Editorial: Gregory L. Clayton

Director of Building Trades:
Taryn Zlatin McKenzie

Executive Editor: Robert Person

Product Manager: Vanessa L. Myers

Editorial Assistant: Nobina Preston

Marketing Director: Beth A. Lutz

Marketing Manager: Marissa Maiella

Marketing Coordinator: Rachael Torres

Production Manager: Sherondra N. Thedford

Senior Content Project Manager:
Stacey Lamodi

Senior Art Director: Benjamin Gleeksman

Library of Congress Control Number: 2011921378

ISBN-13: 978-1-111-03726-0
ISBN-10: 1-111-03726-4

Delmar
5 Maxwell Drive
Clifton Park, NY 12065-2919
USA

Cengage Learning is a leading provider of customized learning solutions with office locations around the globe, including Singapore, the United Kingdom, Australia, Mexico, Brazil, and Japan. Locate your local office at: **international.cengage.com/region**

Cengage Learning products are represented in Canada by Nelson Education, Ltd.
For more learning solutions, please visit our corporate website at www.cengage.com
Visit us at www.InformationDestination.com

NOTICE TO THE READER

Publisher does not warrant or guarantee any of the products described herein or perform any independent analysis in connection with any of the product information contained herein. Publisher does not assume, and expressly disclaims, any obligation to obtain and include information other than that provided to it by the manufacturer. The reader is expressly warned to consider and adopt all safety precautions that might be indicated by the activities described herein and to avoid all potential hazards. By following the instructions contained herein, the reader willingly assumes all risks in connection with such instructions. The publisher makes no representations or warranties of any kind, including but not limited to, the warranties of fitness for particular purpose or merchantability, nor are any such representations implied with respect to the material set forth herein, and the publisher takes no responsibility with respect to such material. The publisher shall not be liable for any special, consequential, or exemplary damages resulting, in whole or part, from the readers' use of, or reliance upon, this material.

Printed in the United States of America
1 2 3 4 5 6 7 14 13 12 11

DEDICATION

To my wife, Diane, for being the love of my life,
my constant inspiration, and for believing
in me through all these years.

ACKNOWLEDGMENTS

I would like to thank the following people who helped make this book possible: To my wife Diane, daughter Erin, and brother Jerry for their contributions; to Mike Feutz, Ph.D., Ferris State University, for writing the Foreword and helping me review my subject matter; to Amy Kavanaugh, Ph.D., Ferris State University, for getting me through my master's degree; to Geoff Moffat for helping me with zoning issues; to Mitch LeClaire and Mike Lafferty for getting me interested in geothermal; as well as the following people and organizations for their support, inspiration, and contributions:

Charles Lacy, Ph.D.; Mary Bigelow; Dave and Sharon Kaechele; Gary and Lois Vanduine; Ryan and Bill Martin; Clayton and Amanda Jackson; Rob Rafson, P.E.; Jeanette Hagen; Tom Lane; Mick Sagrillo; Torresen Marine; Caleffi Hydronic Solutions; John Deere Renewables; and Solar Energy International.

FOREWORD

I first heard of Don Steeby when I began teaching at Ferris State University in 1998. He was the student, I was told, who wrote a software program that our students used to size ground loops for geothermal heat pump systems. His software was successful enough to sell in the market, but it also played a significant role at Ferris. Students in the Heating, Ventilation, and Air Conditioning (HVAC) Engineering Technology program used his software to size ground loops as part of their first-place submissions in a number of international mechanical system selection and design competitions sponsored annually by the American Society of Heating, Refrigeration, and Air Conditioning Engineers (ASHRAE). As you will see, this book is similar to the software that Don wrote. Both are symbolic of his background and career: technical yet down to earth. Don draws on his experiences in life, from the mechanical know-how and connection he made with the environment as a farm boy, to the technical expertise he gained in the HVAC industry and the education he received from a community college and two universities.

Growing up on a dairy farm in west Michigan taught Don to be self-dependent, as his family usually fixed things themselves. Don was milking cows by the time he was 10 and learned about machinery first hand as a young boy, when he found he liked to take things apart to find out how they worked. His interest in agriculture led him to the Institute of Agricultural Technology at Michigan State University. After graduating in 1980, Don found a poor farming economy and turned to other jobs. While working as an electrician he decided that the skilled trades suited him well, so he returned to school part time in the fall of 1985, enrolling in the HVAC program at Grand Rapids Community College. By the time he graduated with his associate degree in 1992, Don had worked his way up to national sales manager, selling gas direct-fired make-up air units for a subsidiary of Rapid Engineering.

Though he had established a successful career, Don wanted to learn more. In the fall of 1994, with a wife, two kids, and a mortgage, he quit his job and went back to school full time to earn his bachelor's degree in HVAC Engineering Technology from Ferris State University. It was at Ferris where Don's passion for learning and down-on-the-farm know-how prompted two of his professors to ask him to write the geothermal software. Upon graduation, his new degree led to a position with the Honeywell Corporation. Controls engineers are called upon to solve complex problems and do so by combining systems thinking with systematic troubleshooting techniques. Because they must know all aspects of countless types of mechanical systems from both the design and operation perspectives, the good ones are among the most knowledge people in the HVAC industry. Don was one of the good ones and moved up through the ranks.

While he was with Honeywell, I asked Don to teach as an adjunct for us at Ferris State. He accepted the new challenge eagerly. He did so well that in 2002, his students won the same ASHRAE international HVAC system design competition that his software had helped other students win in earlier years. Teaching sparked a new interest for Don. In fact, we tried to hire him as a full-time instructor, but he turned us down. Though teaching at his alma mater represented a tempting opportunity, the campus was too far of a drive and strong ties to his community and the family farm did not allow him to consider relocating.

Don also taught as an adjunct at Grand Rapids Community College, another of his alma maters, and located much closer to his home. When a full-time position became available in 2007, he jumped at the opportunity. It was no surprise to me when he was selected as the successful candidate and left a promising job in the HVAC industry to begin a new chapter in his career. True to form, Don took the next step in his education with his new career, and I had the pleasure of serving on his thesis committee when he earned his Master's of Science in Career and Technical Education from Ferris State University in the spring of 2010.

This book is an extension of the research Don completed for his thesis. Much of what you read is a collection of the knowledge that he acquired through his graduate work. But it is more than that. It is a statement of his passion for alternate energy, for learning, and for sharing knowledge. I am reminded of Armstrong International, a family-owned American manufacturer of high-quality products for the steam, air, and hot water industries. Armstrong operates by the motto, "knowledge not shared is energy wasted." That motto has multiple meanings when applied to this book. The energy sources that Don writes of are available and abundant, but like the knowledge in the Armstrong motto, are largely wasted until we deploy methods to harvest them. And the work that Don has done to compile information about the various forms and uses of alternate energy would be wasted if he did not share the knowledge he has gained with you.

This book, written in an easy-to-understand manner, serves as a primer for those who wish to learn about energy alternatives and applications. What follows is a comprehensive work, as Don provides practical, historical, and technical perspectives, allowing the reader to learn about all aspects of each form of alternative energy solutions. This is not an engineering text. It is a thorough introduction to and discussion of solar, wind, geothermal, biomass, and future energy sources (fuel cells and combined heat and power [CHP] systems). Without additional training and/or expertise, the reader should not expect to be prepared to design, install, or maintain an alternative energy system. But the reader will come away with a better understanding of the background, application, feasibility, economics, efficiency, and technology behind these five energy sources.

Don has gone beyond his goal to "outline the fundamental workings of various types of alternative energy equipment and show how these types of equipment are applied, installed, serviced, and maintained for today's marketplace." He has infused his passion and expertise into the pages that follow. From his farm boy know-how to his master's degree in career and technical education, Don has created a resource that you will find informative, practical, and useful.

Michael J. Feutz, Ph.D., LEED AP
Professor, HVAC
Ferris State University

TABLE OF CONTENTS

PREFACE

The dictionary defines the word *energy* as "any source of useable power" and the words *alternative energy* as "energy that can replace or supplement traditional fossil fuel sources." Traditional energy sources such as electricity, natural gas, and fuel oil have been reliable life-sources in society that have created lighting for work, heating for homes, and have made the world a more comfortable place. Today, however, with traditional energy in short supply and in great demand, it is understandable that there is a desire to find alternative sources that will meet the growing needs of our society. The intent of this book is to not only assist the reader in developing a deeper understanding of alternative energy, but also to assist in satisfying the demand and desire for new reference material that has been created by the nation's thirst for clean, abundant energy. The purpose of this book is to enhance the development of practical applications for alternative energy and its equipment within the climate-control industry. There are numerous HVAC installation and service companies throughout the United States who are interested in entering into the alternative energy market. In order to be prepared for this market, these companies will require that their personnel be properly trained in areas of alternative energy in order to become competent and qualified technicians. This book will fill a void that currently exists between basic information on alternative energy and the higher level, more intellectual material that is suited toward the engineering and development of alternative energy systems.

The usage of this book is directed toward the following entities: HVAC contractors and energy contractors who are seeking to educate their workforce in the ways of alternative energy, students who wish to further their education in the use of alternative energy systems, and homeowners and business owners who are seeking alternative ways in which to reduce their energy costs. There is a real need for a comprehensive book that bridges the void between the simplistic, do-it-yourself type manual and the graduate level engineer textbook that tends to focus on the development and analysis of these types of systems. The research that was compiled for the development of this book has resulted in an in-depth study of how these types of systems operate, how they should be properly applied and installed, and how they should be maintained.

The public's desire to develop and utilize sources of alternative energy in the United States will continue to escalate throughout the next several years and beyond. Because of this demand, there will be a perpetual need for trained and qualified technicians who have the ability to understand how to install, commission, service, and repair alternative energy equipment. In order to prepare these technicians, there will be a need for comprehensive information that can be used to train and educate installers, technicians, and service people. Although there is an abundance of individual sources of information regarding alternative energy currently available, there is a need for reference literature that can compile and organize this information into a useable reference book that will meet the needs of today's students, technicians, and building owners.

The use of alternative energy is going to be around for a while. This is not a passing phase. Vast amounts of time and money have been spent investing in the future of alternative energy, and it is very apparent that it will be a viable source for powering America long into the future.

UNIT 1
Solar Energy:
Harnessing the Sun's Power

Chapter 1

INTRODUCTION TO SOLAR ENERGY

One of the most prominent sources of alternative energy is right above our heads. That big ball of light up in the sky that we call the sun is actually a star—a "G2" star to be exact. In reality, it is the most prominent feature in our entire solar system (Figure 1-1). With its surface temperature measuring 11,000°F and consisting of a total mass of more than 99.8% of the entire solar system, the sun generates about 386 billion billion (that is right—*billion billion*) megawatts of power created by nuclear fusion reactions.

At any given time, the sun is delivering up to 1,500 watts of power onto each square meter of the earth's surface that is exposed to sunlight. If you take into account that there are approximately 150 million square kilometers of land surface on the earth, the amount of energy available from the sun is staggering. In fact, more energy from sunlight falls upon the earth's surface in 1 hour than is used by the entire global population in 1 year. Now consider the fact that the United States alone consumed over 3.8 trillion kilowatt-hours of electric power in 2008. For these reasons, it is easy to understand why there is such an interest in harnessing the sun's power as a viable source of alternative energy (Figure 1-2).

Figure 1-1

The sun is the most prominent feature in our entire solar system.

Figure 1-2

Harnessing the sun's power is one of the most viable sources of alternative energy.

Figure 1-3

The effect of the sun's radiant energy as it travels to and from the earth.

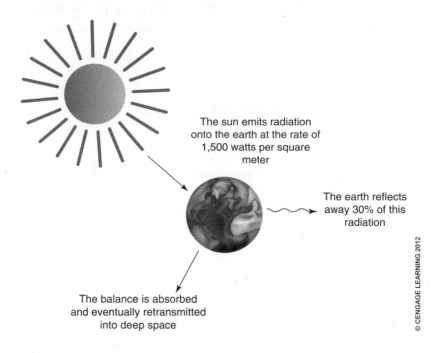

The sun emits radiation onto the earth at the rate of 1,500 watts per square meter

The earth reflects away 30% of this radiation

The balance is absorbed and eventually retransmitted into deep space

© CENGAGE LEARNING 2012

The sun emits visible light and radiation onto the earth. About one-third of this total radiation is reflected back into space. The remaining radiation is absorbed by the earth and then re-radiated into deep space as long-wave infrared radiation (Figure 1-3).

The key to capturing this energy is to allow the radiation to pass through a selective material, such as glass, without allowing it to re-radiate back out. The other method is to directly convert the sun's light into electricity at the atomic level. When we think of the sun as a source for alternative energy, we need to define the two distinct methods by which solar energy is harnessed. The first method is by using the sun's energy to create **thermal storage,** which equates to heating water. This solar thermal storage method can be utilized for a number of applications, including space heating for the home, for the production of domestic hot water, and for the heating of swimming pools and spas (Figure 1-4).

The second method is known as **photovoltaic**. This method involves converting the sun's radiation into useable electrical energy. The electricity that is generated by photovoltaic can be used to diminish or even help eliminate the

Things to Know

The sun is about 4½ billion years old. Light from the sun takes about 8 minutes to reach the earth.

Figure 1-4

Various uses for solar thermal storage.

© CENGAGE LEARNING 2012

dependence that the homeowner or business owner has on the local electrical utility company (Figure 1-5).

Both of these methods will be covered in depth throughout this unit.

A BRIEF HISTORY OF SOLAR THERMAL SYSTEMS

Utilizing the sun's radiation to produce energy is nothing new. In fact, the first commercial solar water heater was introduced in California in 1890. This early solar collector was a simple batch-type water heater (Figure 1-6). It offered hot water that was available in the afternoon and evening, but most of its heat was lost overnight. From then until the 1970s, most thermal storage was limited to a few applications found mostly in Florida.

As a result of the first oil embargo of 1973, and its associated rise in gasoline prices, companies began experimenting with and manufacturing different types of solar energy systems. Many of these earlier systems were not very successful. They were either too complex, had major design flaws, or were just too expensive. However, the systems that were developed in this era laid the groundwork for the explosion in solar manufacturing that would come in the late 1970s. This big

Figure 1-5

A home using a solar photovoltaic system.

Figure 1-6

This solar batch collector is similar to ones used when solar technology was new.

Figure 1-7

Tax credits were a big incentive for solar businesses in the 1970s.

© CENGAGE LEARNING 2012

push into new solar manufacturing was a result of a 40% federal tax credit that was implemented by the U.S. government (Figure 1-7).

This incentive resulted in the creation of hundreds of solar manufacturers and thousands of dealers and contractors rushing to start new businesses. Unfortunately, it also created a number of less-than-reputable marketers who would take advantage of tax credit schemes by selling substandard equipment at inflated prices. These devious distributors and contractors contributed to giving solar products a bad name by selling poorly designed equipment that was often installed incorrectly.

The rush to install solar equipment essentially came to an abrupt end in 1986, when federal tax incentives were discontinued and the price of gas fell below $1.00 per gallon. From then on the public's perception was that the energy crisis was over, and cheap gas ruled the day. It wasn't until just recently solar energy experienced a revival. For instance, the U.S. Energy Policy Act of 2005 implemented a tax credit for consumers who install new solar thermal storage systems. Some of these tax credits will cover up to 30% of the cost of equipment and installation and will not expire until the end of 2016. In addition to tax savings, the efficiency and reliability of solar heating systems has increased dramatically over the past several years. The U.S. Department of Energy (DOE) is working on designing more cost-effective solar thermal storage systems and is also improving the durability of materials that are used in these systems (Figure 1-8).

Figure 1-8

The quality of solar thermal collectors has improved dramatically over the years.

© ISTOCKPHOTO/PAVLO VAKHRUSHEV

Things to Know

The first commercially available solar water heater was called the Climax. It was first introduced by Clarence Kemp around 1890. For an investment of $25, people in California could save about $9.00 per year in the cost of coal.

With other new tax incentives being re-introduced by federal and state governments and with a steady increase in utility costs, solar thermal storage has become one of the most attractive sources of alternative energy available today. More information on government incentives can be found at the Database of State Incentives for Renewables and Efficiency (DSIRE) website (http://dsireusa.org).

THE FEASIBILITY OF SOLAR THERMAL SYSTEMS

Solar thermal storage can conceivably be utilized in any area of the United States. Obviously, there are certain areas that are more conducive to using solar energy as opposed to others. When determining whether solar is the right choice as a means of alternative energy, certain criteria should be followed. Dr. Erich A. Farber was the first individual to be inducted into the National Solar Hall of Fame. He is the former head of the University of Florida's Solar Lab, and a member of the Florida Solar Hall of Fame. Dr. Farber developed five criteria for choosing solar energy over other types of energy in an effort to decide when solar energy is the right choice:

1. **Use only the minimum amount of energy required for the task.** This applies to such principles as correct sizing of the equipment. For instance, do not install a thermal storage system that is designed for eight people when there are only two people who reside in the household.
2. **Use the best source to perform the task.** If the current price that the consumer is paying for electricity is far below the cost to produce it with a photovoltaic panel, then utilizing electricity from the power grid rather than from solar energy only makes sense. Another example would be if a business uses large amounts of **potable** hot water for cleaning and for processing, but also has access to large volumes of waste heat, then solar thermal storage would not be feasible when an alternative free heat source is available.
3. **The equipment has to work correctly.** Oftentimes, it is not the fault of the equipment but rather the fault of the contractor who selected and installed the equipment when malfunction occurs. If a solar panel is inadvertently installed in a shaded area or is undersized and does not meet the required energy demand, the customer will think that solar energy simply does not work. In reality, the equipment probably would function properly if it were applied and installed correctly.

Figure 1-9

Solar energy must fit the needs of the customer.

© ISTOCKPHOTO/ELIANDRIC

4. **The equipment has to be offered at a reasonable price.** The market will determine what the value of the equipment will be. Considering that there are other sources of competitive energy that may be obtained at a lower price, solar thermal energy may need to be marketed as a bona fide return on investment rather than simply as a commodity.

5. **Solar energy needs to fit the social structure and habits of the customer.** This can only be answered by understanding how the household or business conducts its habits and daily routines. For instance, it would not be feasible to install a passive thermal storage system in a manufacturing facility, where most of the energy demands for hot water occur only at night (Figure 1-9).

Following are other regional and site-specific factors that should be considered when determining the feasibility of utilizing energy that is produced from solar thermal storage.

Regional Factors

One regional factor that should be considered when investing in solar thermal storage is the cost of other alternative fuel sources. The heating of domestic water can be accomplished by the use of a number of different fuel sources, such as natural gas, propane, or even wood. The comparison and availability of these types of fuel sources can have a sizable impact on the decision to implement solar energy. In addition, there needs to be a determination as to what the future pricing of these fuels might be. This will factor into the long-term payback of investing in solar energy. If the future price of fossil fuels escalates, the investment payback will be shorter. Current pricing can be obtained by contacting the local utility or supplier of these fuels.

Funding and rebate availability is another factor that can have a considerable impact on the decision to invest in solar thermal energy. As mentioned earlier, the U.S. government is currently offering incentives that will cover up to 30% of the total installed cost on certain solar thermal storage equipment through 2016. These rebates offered by the federal government are in the form of tax credits. More information can be obtained at the website http://energystar.gov. In addition to federal incentives, most states are also offering rebates for qualified installations of solar thermal systems. This information can be researched at the website http://dsireusa.org.

When a solar thermal storage system is incorporated with new construction, the costs can generally be rolled into the initial mortgage or construction loan. When the solar energy project is being used for an existing home or commercial building, the funding for the project can be obtained with the help of such federal agencies as the Federal Home Loan Mortgage Corporation (FHLMC) or through the United States Department of Agriculture (USDA). Further information on federal funding can be obtained through the **National Renewable Energy Laboratory (NREL)** or by researching its website at http://nrel.gov.

The third regional factor that needs to be considered is the amount of **insolation** that is available. Insolation (not to be confused with *insulation*) is the amount of electromagnetic energy in the form of solar radiation that is projected onto the surface of the earth. In short, it is the available sunshine in a given location. This factor affects the efficiency of the solar thermal storage system and is measured by the amount of full sun hours per day. By knowing the insolation level of a given region, the properly sized solar collector that is required can be determined (Figure 1-10).

An area that has poor insolation levels will require a larger collector. Once the insolation level is known for a given region, the correct solar collector can be more accurately sized. Insolation levels are generally expressed in kilowatts per square meter per day ($kWh/m^2/day$). In other words, this number represents the amount of solar energy that strikes a given square meter of the earth's surface in a single day.

Figure 1-10

Solar insolation levels in the United States. Insolation is the amount of energy the sun projects onto the surface of the earth.

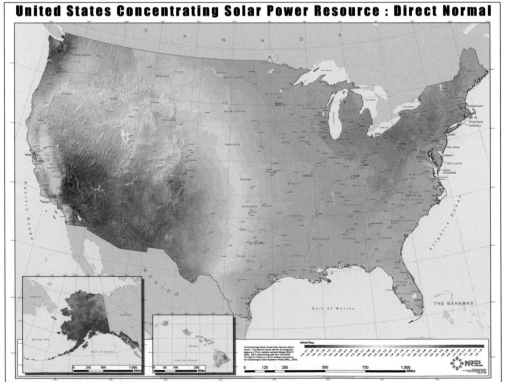

United States Concentrating Solar Power Resource : Direct Normal

COURTESY OF THE U.S. DEPARTMENT OF ENERGY, OFFICE OF ENERGY EFFICIENCY AND RENEWABLE ENERGY

Site-Specific Factors

There are factors that will affect the actual site that is chosen to install a solar thermal storage system. One of these factors is the consideration of a mounting area that faces to the south. This area should be of sufficient size to accommodate a solar collector that is to be mounted on a platform, pole, or on the roof of the building. If the installation is to be roof mounted, the roof must be structurally capable of supporting the entire collector and its associated equipment (Figure 1-11).

It should have accessibility for a service technician should any repairs need to be made. In addition to the location of the solar collector, the mounting area must be free of shading and obstructions. One of the biggest hindrances to the feasibility of the solar collector site is shading from trees and other obstructions. Any shading that casts shadows on the solar collector's area during full-sun hours can drastically reduce the efficiency of the system.

Another site-specific factor is aesthetics. Solar collectors are available in a variety of sizes and with differences in appearance. Any local ordinances that affect the size, location, and appearance of the solar collector should be thoroughly investigated before any decisions are made regarding moving forward with the solar project (Figure 1-12).

Some jurisdictions may dictate what these factors shall be, or may even prohibit the use of roof-mounted solar collectors. Zoning issues that affect solar installations are dependent upon the type of system to be installed. For instance, photovoltaic roof shingles require no additional building permits; however, solar collectors do. Most collectors are considered an "accessory structure" to the building regardless of whether they are roof-mounted, wall-mounted, or free-standing installations.

© ISTOCKPHOTO/ PAVLO VAKHRUSHEV

Figure 1-11
Roofs must be structurally capable of supporting solar collectors.

Figure 1-12

Local ordinances may have restrictions on roof-mounted solar collectors.

A9PHOTO, 2011. USED UNDER LICENSE FROM SHUTTERSTOCK.COM

Green Tip

On Zoning Laws and Solar Panels

Green energy alternatives such as solar panels have local municipalities scrambling to keep up, as they must come up with guidelines to regulate privately owned alternative energy projects for both residential and commercial properties. Some states, such as California, already have laws that are designed to protect the consumer's right to install and operate solar energy systems on homes and businesses. Hawaii is the first state to require that all new home constructions must have solar water heaters. In fact, beginning in 2010, all new single-family homes must include this attribute in order to obtain a building permit. However, the main focus of most zoning regulations pertains to the visibility of solar panels from an aesthetic viewpoint, and most do not allow for free-standing solar panels in residential areas. Still other zoning laws and municipal ordinances must address such realistic concerns as:

- The exceeding of roof loads
- The use of unacceptable heat exchangers
- The improper wiring of solar equipment
- The unlawful tampering with potable water supplies

What can be done if there are no specific ordinances or zoning laws in a particular city or township where there is a desire to install solar equipment? First, contact the local township, village, or city office and speak with the zoning administrator. Find out if there are any regulations on solar panel installation and what those restrictions might be. Attend township or city board meetings and request that there be positive legislation developed that protects both the consumer and neighborhood in order to advance the use of alternative energy projects. Maintain a positive attitude when working with city or township officials; they are there to serve the community, not create enemies.

In most instances, the local municipality will require either a site plan for the solar installation if it is to be free standing, or a building plan if it is to be roof mounted. Shop drawings that include the manufacturer's specifications on the solar collector and its accessories will need to be included as part of the site plan or building plan.

ECONOMICS AND PERFORMANCE OF SOLAR THERMAL SYSTEMS

The economics of a solar thermal storage system are dependent upon a number of factors. These factors include: the amount of hot water that is used, the performance of the solar thermal system, the geographic location of the solar collector, the comparative cost of conventional fuels, and the cost of the fuel being used as a backup source for the heating system. Solar thermal storage systems are most cost-effective when they are utilized for the majority of the year, such as in cold climates where the solar resources are plentiful. They are the most economical when they displace more expensive heating fuels, such as propane or electricity. For instance, when the local electrical rate exceeds $0.07 per kilowatt-hour or the cost of propane is above $1.25 per gallon, solar thermal energy can be an attractive investment. At these rates, the average household could save between 50% and 80% of the total cost to heat domestic water. These savings can increase if the system is used for other applications, such as the heating of a pool or spa.

The installed cost of an average solar thermal system will vary, but most commercial systems will range from $30 to $80 per square foot of the collector area. Residential systems are usually slightly lower in cost. Typically, the larger the system, the less it costs per unit of collector area. The average home will usually need approximately 100 to 120 square feet of solar collector for domestic hot water needs. If the system is being financed through new

Things to Know

The use of solar heating panels for swimming pools has grown to the number-one application of solar energy in the United States today. When properly sized and installed, a solar heating system for use with a swimming pool can pay for itself with energy savings in 2 to 3 years.

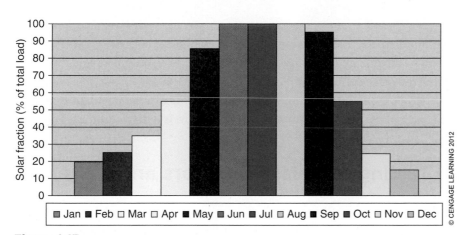

Figure 1-13
The solar heat fraction of a system is equal to the amount of energy that the solar thermal system displaces over conventional energy.

© CENGAGE LEARNING 2012

construction, such as through a mortgage or through refinancing, the initial cost can become even more attractive. On a new 30-year loan, the cost of an installed system including the solar water heater is usually between $13 and $20 per month.

One way of expressing the thermal performance of an average solar energy system is by the solar heating fraction (Figure 1-13). This expression is equal to the amount of energy that the solar thermal system displaces over the conventional energy consumption and can be measured on a monthly or annual basis.

For instance, if the solar thermal storage system has an annual solar heating fraction of 0.15 for the month of January, it would mean that the system replaced 15% of the normal energy costs for that month. On an annual basis, the solar heating fraction can range from 41% in a predominantly cloudy location to as much as 94% in an area where sunshine prevails year round. Actual savings will depend upon a number of factors, including:

- The size of the solar collector and storage tank
- The average temperature of the cold water supply
- The amount of sunlight in a given region
- The average volume of monthly water usage

Because of the unlimited combinations of solar collectors, storage tanks, temperature controls, and many other factors, it is essential that there be specialized software available as a tool for evaluating the technical performance of an active solar storage system. Given the complex nature of the solar energy available for a given area, a software program that simulates the performance of various systems in specific geographic regions is essential.

Things to Know

A typical residential solar water heating system for a family of four delivers about 4 kilowatts of electrical-equivalent thermal power when under full sun, when the temperature of the water in the storage tank is about the same as the air temperature. A system such as this will require about 64 square feet of solar collector surface area.

Fortunately, there are several software programs available on the market that will perform this task. The efficiency of a given system depends upon the adequacy of the storage capacity, the solar collector and its orientation to the sun, as well as the quality of the system's components, appropriate insulation, heat exchanger design and efficiency, operational cost, and the time of day that there is a demand for hot water.

Certification and Testing of Solar Collectors and Systems

Most of today's modern solar thermal storage equipment is certified by a nationally recognized testing agency. The Solar Rating and Certification Corporation (SRCC) was created in 1980 as a nonprofit organization whose primary purpose is the development and implementation of certification programs and national rating standards for solar energy equipment. The SRCC administers a certification, rating, and labeling program for solar collectors.

It also has a similar program for complete solar water heating systems. Independent laboratories that are accredited by the SRCC administer nationally accepted performance tests on solar equipment as a means of maintaining nationally recognized standards and guidelines. These test results and product data are then evaluated by the SRCC to determine the product's compliance with the minimum standards for certification and to calculate the equipment's performance ratings. Any equipment that has been certified and rated by the Solar Rating and Certification Corporation is required to bear the SRCC certification label, which shows the performance rating for that product (Figure 1-14). In addition, each certified product is published in the SRCC directory. The directory listing for each product contains information on the product's material and specifications as well as the certified performance rating.

One of the benefits that solar equipment manufacturers receive as a result of obtaining SRCC certification is a reliable means of judging the product's durability and performance on a standardized basis. The certification also gives manufacturers national recognition, and they are only required to certify the product once. The contractors who are installing SRCC-certified solar equipment benefit by offering product credibility, and a defense against unethical competition and false claims (Figure 1-15).

Figure 1-14

Testing solar thermal storage equipment. Product data are evaluated by the SRCC to determine compliance with minimum standards for certification.

Figure 1-15

Contractors benefit by offering products certified by the Solar Rating and Certification Corporation.

Green Tip

On How Much Can a Solar Panel Save?

The average family of four uses approximately 70 gallons of hot water per day. The standard water heater typically raises the temperature of the water from 50°F to 120°F, which equates to a 70°F temperature rise.

A gallon of water weighs 8.34 pounds; therefore the solar thermal system would need to heat 584 pounds of water per day. By definition, 1 BTU (British thermal unit) is the amount of energy required to raise 1 pound of water 1 degree Fahrenheit, so it would take 40,866 BTUs of energy per day to meet this demand (584 pounds times 70-degree temperature rise).

The price of various heating fuels can be compared by calculating the cost per million BTUs of each type of fuel. For instance, if the price of electricity is $0.10 per kWh, the cost per million BTUs is $29.33. Conversely, if the price of natural gas is $1.30 per hundred cubic feet, its cost per million BTUs is $13.00.

At these rates, it would cost $1.20 per day or $35.95 per month to heat the water with electricity, and $0.53 per day or $15.90 per month to heat with natural gas. If a solar collector that is sized between 80 and 100 square feet is used to meet the hot water requirement used in this example, the annual cost savings would be $431.40 over electricity and $190.80 over natural gas.

The consumer of solar equipment also receives benefits from SRCC programs. These benefits include obtaining a measurement of quality and performance, third-party independent testing, and a nationally standardized method of comparing other solar equipment that allows them to determining their best choice before purchasing. Finally, the federal and state governments benefit from SRCC programs by having a rational basis for tax credit evaluation and qualifying regulations as well as having a basis for setting codes and standards. The Solar Rating and Certification Corporation provides various documents that are available to managers, building owners, and homeowners alike who wish to implement solar energy programs.

2

HOW THERMAL STORAGE WORKS

Thermal storage is achieved by capturing the sun's energy in a solar collector by allowing its radiation to pass through a selective material without allowing it to re-radiate back out. The collector contains one of several different types of piping through which water or air is circulated. A number of different thermal storage configurations can be utilized. The most appropriate type depends upon certain factors, including the geographic location, the amount of hot water or air that needs to be generated, and the specific heating or cooling application to be used. The following is an analysis of each type of system in detail.

PASSIVE SYSTEMS

A passive solar heating system harvests energy from the sun without the use of electrical or mechanical devices. In its simplest form, this type of solar heating system can constitute a home with south-facing rooms and windows that has an open-space floor plan to optimize thermal mass (Figure 2-1).

This type of application is usually implemented during the home's design and construction phase and incorporates such facets as the orientation of the home in relation to the sun and the types of construction materials that will best be suited for maximum solar gains.

Figure 2-1

A house utilizing passive solar heating typically includes south-facing windows.

Integral Collector Storage Unit (ICS)

A passive solar thermal storage system consists of a solar collector and some type of solar heating water storage device. One example of this type of storage system is an **intergral collector storage (ICS)** unit. With an ICS unit, a 30- to 40-gallon storage tank is painted flat black and mounted in an insulated collector box that has glazing (glass) on the side exposed to the sun (Figure 2-2).

The collector box is usually lined with aluminum-foil-faced foam insulation that is sloped and curved toward the tank. The best installation design for this type of system is to use a single tank in the box and mount it horizontally with the tank facing from east to west. The curved reflector surfaces are angled to reflect the sunlight onto the tank (Figure 2-3).

When the sun shines through the glass, its radiation is absorbed by the tank and in turn the water is heated. With this type of system, the water itself is the solar collector. In most ICS units, the hot water is typically drawn from an outlet at the top of the tank, where it is the hottest, and cold water passes out through the bottom of the collector. It is then piped indoors to the main water heater, a backup water heater, or a solar thermal storage tank.

One of the advantages of this type of system is the fact that it has a relatively simple design; therefore the installation costs are low. In addition, ICS units do

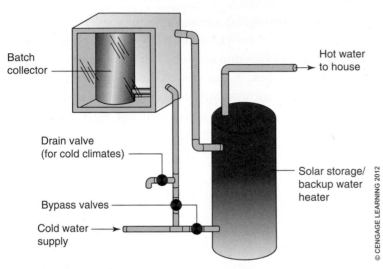

Figure 2-2

An intergral collector storage (ICS) device consists of a solar collector and storage tank.

© CENGAGE LEARNING 2012

Figure 2-3

A foil-lined collector box mounted horizontally from east to west.

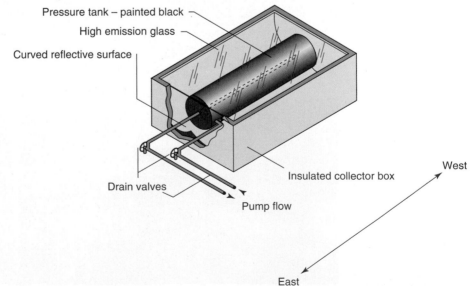

© CENGAGE LEARNING 2012

not require a great amount of maintenance because there are no moving parts. Furthermore, because no electrical devices are being used to transfer water, there are essentially no operational costs. There are, however, several disadvantages to this type of passive thermal storage system. One disadvantage is that in colder climates the water in the tank will cool down overnight—as much as 30°F to 40°F below the supply water temperature. Therefore, hot water is available only during the daytime hours when there is a sufficient amount of direct sunlight on the tank. The optimal usage period for domestic hot water is typically between the hours of 12 p.m. and 8 p.m. Furthermore, certain areas of the United States should avoid using this type of system, especially where the nighttime temperature is below 60°F from late fall to early spring. This is due to the extreme overnight heat loss. In these areas, the ICS should be used on a seasonal basis or not at all. Another cautionary measure that needs to be taken into consideration is the tank weight. An empty tank itself can weigh up to 250 pounds. If it is filled with 30 to 40 gallons of water, the total weight can easily exceed 500 pounds. Therefore, consideration needs to be taken for proper rigging and for correctly mounted roof supports. Finally, it is essential that the proper type of glass be used. The glazing on the exterior of the box should be low-iron tempered glass—never plastic, Teflon, acrylic, or fiberglass.

Thermosyphon Systems

Some solar thermal storage units incorporate a sloped solar collector that is installed in conjunction with an insulated tank. The tank may be mounted either indoors or outdoors. This type of system is referred to as a **thermosyphon** solar hot water heating system (Figure 2-4).

Like ICS units, thermosyphon systems do not incorporate pumps or other controls. They utilize the principle of natural convection, which is that hot water rises and cold water falls. When the solar collector is heated by the sun's radiation, the water in the collector expands. Because the heated water has a lower density than the cooler water located at the bottom of the storage tank, a gentle circulation effect takes place. The warmer water rises to the top of the collector and into the tank, where it is stored. This is why the tank is always located above the collector. Because the cooler water is heavier, it descends to the bottom of the collector. This process is known as thermosyphoning and takes place without the need of a circulating pump (Figure 2-5).

© ISTOCKPHOTO/GEORGIOS ALEXANDRIS

Figure 2-4

A thermosyphon solar thermal collector.

Figure 2-5

A thermosyphon system uses natural convection to circulate the water.

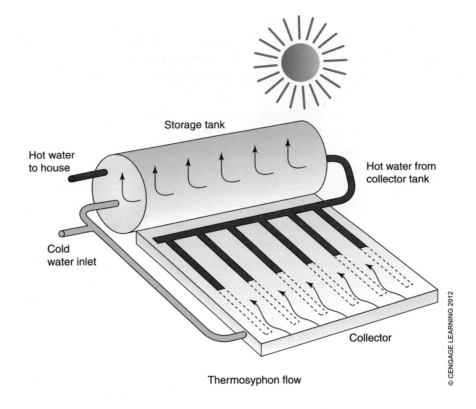

Thermosyphon flow

© CENGAGE LEARNING 2012

Figure 2-6

A piping diagram for a thermosyphon system.

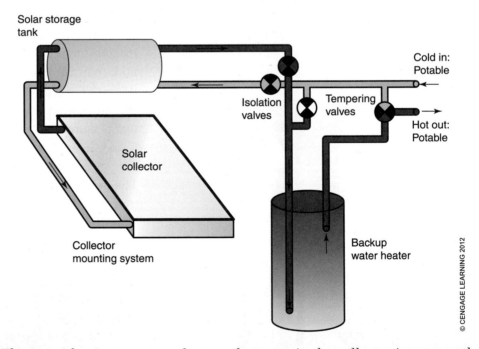

© CENGAGE LEARNING 2012

Thermosyphoning occurs as long as the water in the collector is warmer than the water in the storage tank. As the sun's radiation diminishes, the water in the collector cools to a temperature below that of the storage tank, and the water's natural flow ceases. The water from the storage tank is piped directly into the water heater located inside the home (Figure 2-6). Certain precautions should be taken with this type of system because extreme water temperatures are possible during prolonged periods of sunny weather. To prevent scalding conditions, a thermostatic mixing valve can be installed on the outlet of the water heater. This valve is sometimes called a tempering valve and automatically mixes a portion of the cold supply water with the hot water to maintain the desired water temperature delivered to the home.

Tech Tip

Recommended Hot Water Temperatures

Most experts recommend that the temperature of domestic hot water should not exceed 125°F. The International Plumbing Code states that the maximum hot water temperature for a shower or bathtub shall be 120°F. Water temperatures exceeding this setpoint can pose a serious risk of scalding burns, particularly to children. However, in order to sanitize dishes, a hot water temperature of 140°F is required. There are some concerns that lowering the water temperature will result in soap working improperly in dishwashers or washing machines. Actually, most soaps and detergents are designed to work best at temperatures between 120°F and 125°F.

Thermosyphon systems are relatively inexpensive, and because there are no electrical or mechanical controls as on the ICS system, they are quite reliable and pose few maintenance problems. Contrary to the ICS system, thermosyphon systems incorporate an insulated tank and therefore maintain their water temperature many hours after the sun goes down. However, like the ICS system, thermosyphon systems rely on the outdoor air temperature in order to function properly. This means that they will stop working if the outdoor air temperature falls below freezing. For this reason, thermosyphon systems are best suited for warmer climates or for seasonal usage only. In addition to being temperature sensitive, as in the ICS system, storage tanks are quite heavy and usually require special roof reinforcement.

ACTIVE SYSTEMS

What separates an **active system** from a passive solar thermal storage system is the incorporation of a mechanical means of fluid transfer. For instance, an active system that is used primarily for space heating would implement the use of a fan to circulate air through the system. This type of configuration may use an **inline fan** that is ducted to the solar collector, such as is shown in Figure 2-7, or may simply use a conventional ceiling fan to circulate the room air (Figure 2-8).

Most active systems, however, incorporate the use of an energy storage system that will provide heat when the sun is not shining. These types of systems are typically liquid based and use either water or an antifreeze solution that is circulated through a **hydronic collector** by means of a pump. They can be used for space heating applications, for heating of domestic water, or for a combination of both.

Figure 2-7

An inline fan ducted to a solar collector can act as an active solar system.

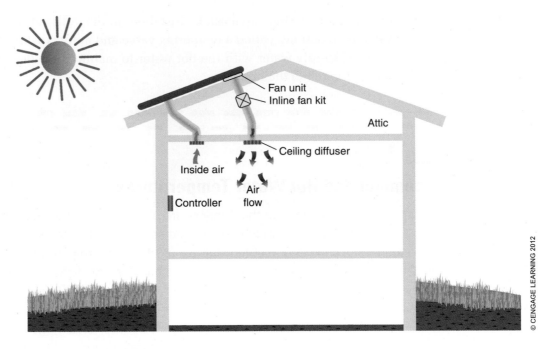

Figure 2-8

Use of a ceiling fan to circulate the room air can also act as an active solar system.

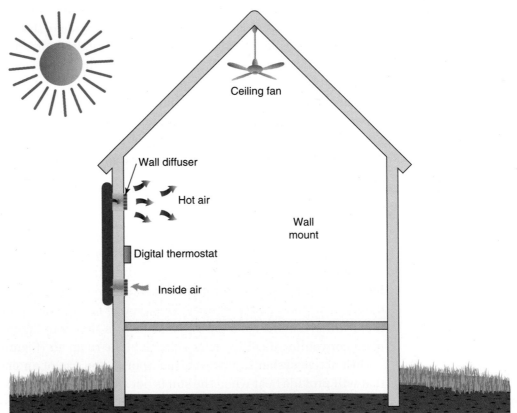

Numerous different configurations are used when incorporating an active solar thermal storage system. These configurations include:

- Open-loop systems
- Closed-loop systems
- Drainback systems
- Pressurized systems
- Unpressurized systems

Open-Loop Systems

An **open-loop system** is one where the hot water is piped from the solar collector directly into a storage tank or a domestic water heater (Figure 2-9). The system is called an open loop because the piping from the collectors to the storage tank or water heater is open to either the home's well water supply or to the city water system. With this type of system, a pump is used to circulate the water from the solar collector to the storage tank or water heater. In some applications, a primary storage tank may be used in conjunction with a backup water heater.

Figure 2-9

An open-loop thermal solar system using two storage tanks.

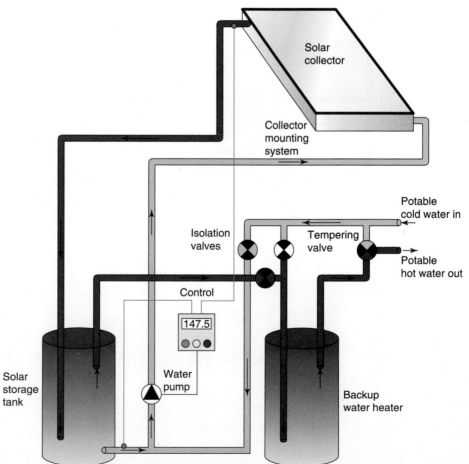

The water flow through the collector is controlled by measuring the differential temperature between the water in the storage tank and at the outlet of the collector. When the water temperature at the outlet of the collector is warmer than in the tank, a controller energizes the pump, which circulates the water through the storage tanks and back to the solar collector.

The open-loop thermal storage system is very efficient, simple to install, and quite reliable. The circulator pump that is used can be as small as 10 watts. If the pump runs off **direct-current (DC)** voltage, it can be powered directly by a photovoltaic module. However, there are several drawbacks to open-type systems. Because they utilize the domestic water supply, open-type systems cannot circulate an antifreeze solution through them, and therefore they are subject to freezing. If these systems are used in areas where freezing temperatures are experienced, they will need to be drained whenever the outside air temperature is expected to fall below 35°F. In addition to freezing conditions, open-type solar thermal systems are subject to the quality of the water being used. Water that is acidic or contains a high level of rust or dissolved solids (hardness) should be avoided. Hard or acidic water will corrode the copper piping or develop scale buildup and lead to premature failure of the collector.

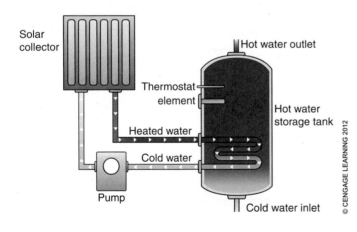

© CENGAGE LEARNING 2012

Figure 2-10

A closed-loop thermal solar system.

Closed-Loop Systems

A **closed-loop system** is similar to an open-loop system, except that it circulates a heat-transfer fluid through the solar collector and through a closed heat exchanger inside of the storage tank (Figure 2-10).

The heat-transfer fluid that is used usually consists of a mixture of **glycol** and water. This type of mixture is necessary in any climate where there is a potential for freezing temperatures. The glycol that is used in these systems is usually food-grade **propylene glycol.** The glycol acts as an antifreeze for the fluid

Things to Know

PROPYLENE GLYCOL AND THE ENVIRONMENT

Propylene glycol is a common additive in beverages. It is biodegradable and will not concentrate in common water systems. Furthermore, the effects on aquatic organisms have been shown to be practically nontoxic.

Figure 2-11

When using solar collectors for thermal storage in cold climates, freeze protection must be included.

© ISTOCKPHOTO/SCHMITZ OLAF

and is nontoxic in the event that it should ever come in contact with the domestic water supply. A single night of subfreezing temperatures can severely damage a solar collector that does not utilize some type of antifreeze. This situation can occur even in traditionally warmer locations such as Arizona, Texas, and Florida. This is why all closed-loop thermal storage systems used in the United States and Canada should employ some method of freeze protection (Figure 2-11).

Conversely, solar thermal storage systems that use glycol should never be designed to operate at over 195°F on a continuous basis. Doing so will cause the glycol solution to degrade and quickly turn to **glycolic acid**. This will cause copper piping to corrode and eventually fail. In addition, glycol-filled systems should not be allowed to remain idle for a long period of time. If the solution is not circulated on a regular basis it can become stagnant and form sludge and organic acids, especially during hot, sunny weather. This situation will also cause corrosion in the piping and in the collector, as well as substantially reduce the rate of heat transfer. Problems with freezing and glycol stagnation can be prevented by incorporating either a drainback system or a drain-down system into the thermal storage system, as described next.

Drainback Systems

Drainback systems offer a viable alternative method of freeze protection for solar thermal storage systems and can be safely installed anywhere in the United States. This method works by simply draining all of the water or glycol solution from the solar collector and any exposed piping into a drainback tank whenever the system is not collecting solar energy. The drainback tank is located inside the building, where it is protected from the outside elements (Figure 2-12).

The drainback system relies upon gravity along with properly pitched piping to quickly drain the water or glycol solution when the system's circulator pump shuts off. When the system is initially filled with water, the drainback tank is only filled to a predetermined level. When the circulating pump is de-energized and the

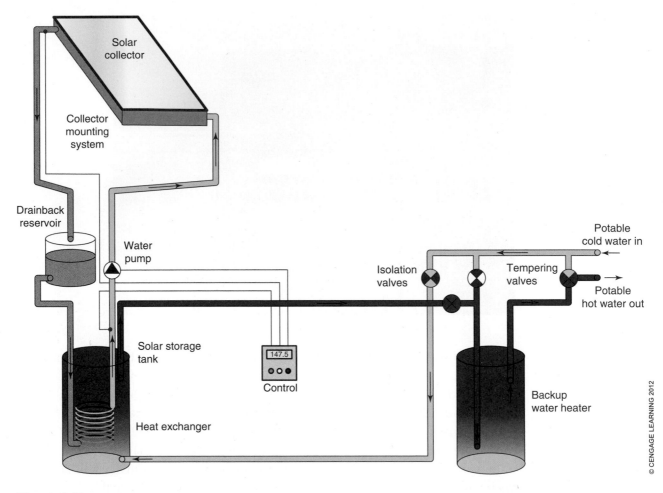

Figure 2-12

A solar thermal storage system using the drainback method.

system is drained back, all of the piping above the tank's water level, as well as the piping in the solar collector, is displaced with air. This prevents any damage to the exposed piping when the outside air temperature drops below freezing. All of the piping and other components below the drain tank's fill level are filled with liquid. When the system circulating pump is energized, liquid is pumped up into the collector. By doing so, air is forced ahead of the liquid, displacing it and eventually returning to the drainback tank. This process causes a slight drop in the drainback tank's liquid level because it replaces the air in the solar collector and system piping. One item that should be taken into consideration when incorporating a drainback system is that **air vents** should not be used (Figure 2-13). See "GreenTips" on the use of air vents on solar collectors.

Pressurized and Unpressurized Systems

A pressurized solar thermal storage system is one where the circulation loop is closed to atmospheric pressure. This type of system is characteristic of the closed-loop and drainback-type systems. An **unpressurized system** is one where the loop

Figure 2-13

Automatic air vents must be used on a closed-loop solar collector.

COURTESY OF DONALD STEEBY

is open to the atmosphere, such as with an open-loop system. With any of the above-mentioned loop configurations, proper piping practices should be adhered to. These practices include the use of an **expansion tank**, pressure-regulating valve, water-regulating valve, high-temperature limit, and **pressure-relief valve**, as well as proper placement of the circulation pump.

SOLAR COLLECTORS

Solar collectors act as the key component of the thermal storage system. They gather the sun's radiant energy and transform the radiation into heat. This energy is then transferred to the system's heat-transfer fluid, which consists of either water or

Green Tip

Air Vents on Solar Collectors

An air vent is a device that is installed at the highest point on the piping loop, usually at the outlet of the collector. When laying out the piping arrangement for a solar thermal storage system, it is important to know which type of system needs an air vent. The most common type of system that would incorporate the use of an air vent is a closed-loop pressurized system. With this type of system, it is important to eliminate air that may become trapped in the piping or in the collector itself. Trapped air can cause a number of problems, including poor or no heat transfer because of a lack of circulated heat-transfer fluid, as well as corrosion in the piping and undesirable noise.

glycol. Nearly all liquid-based solar thermal storage systems use one of two types of solar collectors:

- **Flat-plate collectors**
- **Evacuated-tube collectors**

Most residential and commercial applications for solar thermal storage that require the fluid temperatures to be below 200°F will typically use flat-plate collectors. Those applications requiring liquid temperatures higher than 200°F will usually implement the use of evacuated-tube collectors.

Flat-Plate Collectors

Flat-plate collectors are the most common type of solar collector used with thermal storage systems (Figure 2-14). A typical flat-plate collector consists of an insulated metal or aluminum box covered with tempered low-iron glass, called glazing. This glazing can withstand high thermal stress as well as hazards such as hailstones. It utilizes a low iron-oxide content to minimize the absorption of solar radiation as the sun's rays pass through it.

The main component of the flat-plate collector is the absorber plate. This plate is usually an assembly of copper sheeting with copper tubing fastened to it. The top of the absorber plate is typically coated with a dark-colored paint or with a special coating that readily absorbs radiation as the sun passes through the glazing and strikes its surface. As the absorber plate heats up, the sun's energy is transferred to the fluid that is circulating through the copper tubing that is connected to the plate. This takes place because the fluid is cooler than the absorber plate. As the fluid absorbs heat, it is pumped through the collector and to the storage tank or heat exchanger (Figure 2-15).

Figure 2-14

A flat-plate solar collector.

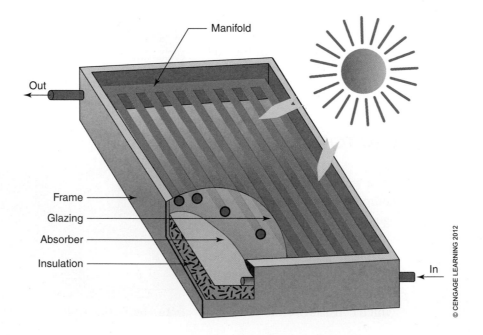

Figure 2-15

A flat-plate collector used to heat domestic water and for comfort heating.

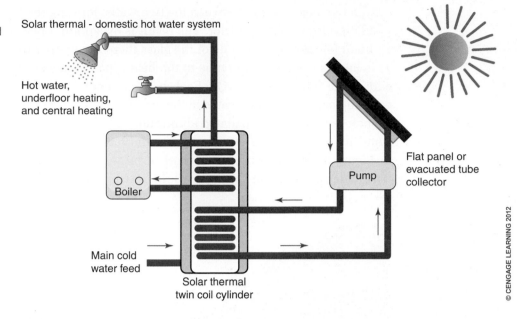

Solar thermal - domestic hot water system

Hot water, underfloor heating, and central heating

Main cold water feed

Boiler

Solar thermal twin coil cylinder

Pump

Flat panel or evacuated tube collector

© CENGAGE LEARNING 2012

Figure 2-16

Roof-mounted flat-plate collectors typically used for the production of hot water.

© ISTOCKPHOTO/START8P

In areas where there is an average amount of solar energy available, flat-plate collectors are usually sized for approximately 1 square foot per gallon of hot water for 1 day's use (Figure 2-16).

Evacuated-Tube Collectors

This type of solar collector consists of a row of glass tubes, each having concentric-shaped inner and outer walls. The air between these walls has been removed, forming a vacuum. This vacuum essentially eliminates any convective and conductive heat transfer between the two walls, forming the best possible thermal insulation for a solar collector. The result is exceptional performance even at low **ambient** temperatures. Inside of the glass there is a copper tube that absorbs heat. This heat is carried up the tube to the heat pipe condenser located outside of the end of the tube (Figure 2-17). Each heat pipe condenser is then connected to a common copper header located at the top of the collector (Figure 2-18). The header is enclosed within an insulated aluminum manifold, where the heated fluid flows from the collector to a storage tank or heat exchanger (Figure 2-19). As shown in Figure 2-20, solar energy is absorbed by the inner heat pipe and is carried to the condenser, where it warms the fluid flowing through the manifold. The roof- or

Figure 2-17

The inside of an evacuated solar collection tube.

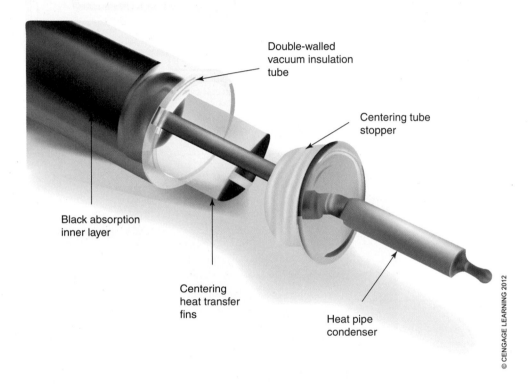

Figure 2-18

Evacuated tubes are connected to a copper header.

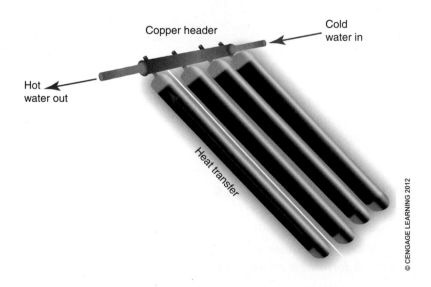

Figure 2-19

An aluminum manifold housing the main copper header.

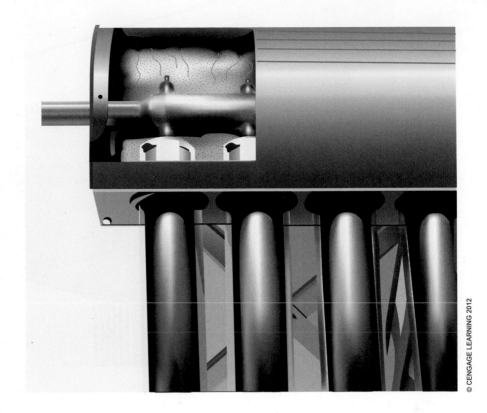

© CENGAGE LEARNING 2012

Figure 2-20

Heat transfer is achieved through the evacuated-tube collector.

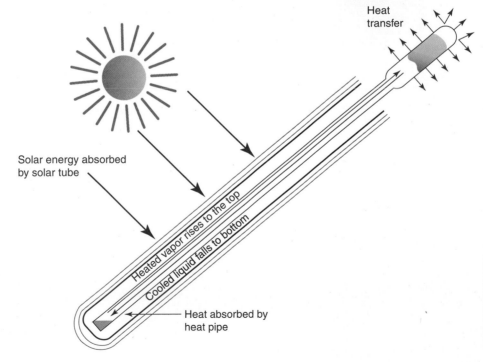

Heat transfer

Solar energy absorbed by solar tube

Heated vapor rises to the top

Cooled liquid falls to bottom

Heat absorbed by heat pipe

© CENGAGE LEARNING 2012

Figure 2-21

A group of wall-mounted solar evacuation tubes is known as an array.

COURTESY OF DONALD STEEBY

Field Tip

Solar Evacuated Tubes

Because they use radiation rather than convection for heat transfer, evacuated-tube collectors will not heat up like flat-plate collectors. Thus in the winter, they will not melt large quantities of snow that fall on them at one time. It can be very difficult to clear the snow from the glass tubes without breakage, so be careful.

wall-mounted evacuated tube collectors are collectively known as an array (Figure 2-21).

The decision on which type of solar thermal storage system to invest in will depend upon a number of factors, including the geographic location, the amount of hot water that is necessary to generate, and the type of heating fuel available for auxiliary use. When making a large investment in solar energy, it is best to consult with local distributors and other customers who have purchased similar systems. This will ensure that the best purchases will be made based upon sound decision making and comparative analysis.

APPLICATIONS FOR SOLAR THERMAL STORAGE

Chapter

3

Once the decision has been made regarding which type of solar thermal storage configuration to choose, the next step is to properly install the equipment based upon the application to be utilized. There are several excellent uses for solar thermal storage systems, and these uses can be suited for both residential and commercial applications.

It is important to remember that the use and application of solar thermal energy involves three steps: It must first be collected, it must be stored, and then it must be distributed. This is with the understanding that the medium being used to transfer this solar energy is water, or a water/antifreeze solution. Collection of this energy takes place at the outdoor solar collector, which has been discussed previously in Chapter 2. The storage of solar energy involves the use of some type of tank or vessel, which has also been covered in Chapter 2. The distribution of the sun's energy can be accomplished through the building's existing plumbing system, either through the hot water system or central heating system, or it may be distributed through a dedicated solar thermal system, which includes a storage tank.

The uses for solar thermal energy can be applied for individual heating needs or a combination of several applications. For residential applications, these uses include: swimming pools, hot tubs and spas, space heating, and, of course, domestic hot water (Figure 3-1).

For commercial applications, the use of solar thermal energy can be applied to food service areas such as restaurants and bakeries, commercial car washes, and for the hot water needs of multi-unit buildings such as apartments and condominiums (Figure 3-2).

There are now even applications available for solar-powered air conditioning.

Figure 3-1

Solar thermal energy can be used to heat swimming pools as well as spas and hot tubs.

© ISTOCKPHOTO/RADOSLAW KOSTKA

© ISTOCKPHOTO/STU SALMON

Figure 3-2

Car washes and high-demand domestic water usage, such as with condominium projects, are a good application for commercial solar panels.

SOLAR THERMAL SYSTEM INSTALLATION

Before the equipment can be installed, the proper placement of the solar collector must be determined. In order to accurately determine the placement, a site survey is necessary. Before an accurate site survey can be accomplished, one must have an understanding of the relationship between the earth and the angle of the sun's rays.

Solar Angles

The earth revolves around the sun on an axis that passes through the north and south poles. This axis is tilted at an angle of 23.44° with respect to the earth's orbital plane, which is also known as the declination angle (Figure 3-3).

The tilt of the earth explains why there are changes of seasons and changes in the amount of daylight as the earth makes its annual orbit around the sun. The intensity of the sun's radiation is also significantly affected, and this effect is observed as the sun's path changes across the sky. The sun's exact position in the sky can be determined by the measurement of two distinct angles. The first

Figure 3-3

The earth revolves around the sun on an axis angle of 23.44°.

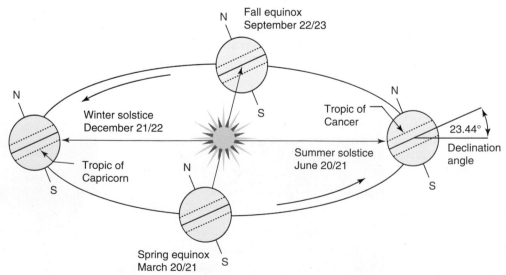

angle is known as the **solar altitude angle** and is measured from the horizontal surface of the earth to the center of the sun. The second angle is the solar azimuth. This angle is measured starting from true north, which is considered 0°, and travels in a clockwise direction until it intersects the position of the sun (Figure 3-4).

Figure 3-4

The azimuth and altitude angles of the sun in relationship to the earth.

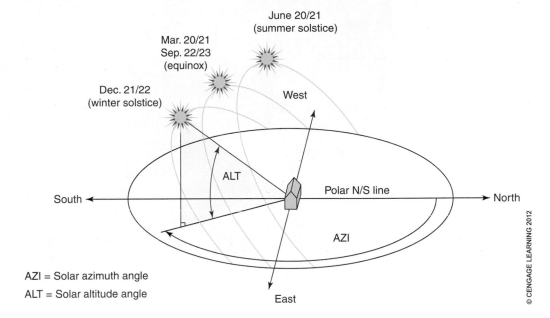

AZI = Solar azimuth angle
ALT = Solar altitude angle

© CENGAGE LEARNING 2012

These two angles are in constant variation as the sun travels across the sky, and also vary according to different latitudes and longitudes. The calculation of these angles will play an important role in determining where the solar collector will be mounted. Fortunately, these two angles have been precisely measured, and can be calculated for any time and location anywhere on the face of the earth. As a convenience, the University of Oregon Solar Monitoring Laboratory website can generate a solar path diagram for any location and time at http://solardat.uoregon.edu/SunChartProgram.html.

Solar Collector Panel Positioning

Proper positioning of the solar collector will greatly affect the performance and longevity of the solar thermal storage system. One of the biggest issues with regard to site selection is the amount of shading that may interfere with the collector. Obviously, the amount of shading from obstacles such as trees, hills, nearby buildings, or other objects must be held to a minimum. One device that is used to determine the effects of shading is called a solar pathfinder. This device is placed at the location where the amount of shading is to be evaluated. After the pathfinder has been placed in the proper orientation, its clear hemispherical dome projects the reflections of nearby objects onto a special chart that will indicate when the desired location is in the shade (Figure 3-5).

Preferably, the solar collector should be unshaded for as long as the sun is shining. However, as a rule, no portion of the collector should be shaded between the hours of 9:00 a.m. and 3:00 p.m. real solar time for every day of the year.

Figure 3-5

A solar pathfinder is used to determine the effects of shading on the solar panels.

The proper angle at which the solar collector is mounted is also of importance. Collectors that are mounted facing due south will receive the optimum amount of insolation (solar radiation). In the northern hemisphere, due south is equivalent to the azimuth angle being 180°. In some instances, because of the orientation of a given building, this exact angle may not be achievable. Fortunately, the total annual solar energy captured by the collector is not completely subject to the azimuth angle. Variations of up to 30° east or west of true south will only reduce the amount of annual solar energy collected by about 2.5%.

The other important factor when determining site selection and mounting is the slope angle of the solar collector. The ideal angle of slope will depend upon the latitude of the geographic location as well as the intended application of the system. For instance, collectors that are used for domestic water heating should be sloped at an angle that is equal to the local latitude. However, variations of $+/-$ 10° on this angle will not greatly affect the performance of total solar energy that is collected. If the system is being used for a space heating application, the collector should be at a steeper angle to take advantage of the sun's angle during fall, winter, and spring (Figure 3-6).

For this application, use the local latitude and add 10 to 20° to the slope. Using this steeper slope will actually reduce the amount of solar collection that takes place during the summertime. However, it will prevent the overheating of larger solar arrays that are often used for space-heating applications. Even though this practice may appear counterproductive, these larger arrays often provide a large percentage of the solar energy needed during warmer weather. For dual applications of both space and domestic water heating, a slope angle of the local latitude

Figure 3-6

If the solar collector is being used for a space heating application, it should be at a steeper angle to take advantage of the sun's angle during fall, winter, and spring.

© ISTOCKPHOTO/UKO JESITA

Figure 3-7

Examples of various collector panel locations to ensure that they are facing due south and allow for the proper slope.

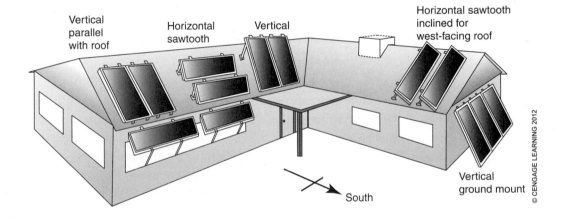

Vertical parallel with roof

Horizontal sawtooth

Vertical

Horizontal sawtooth inclined for west-facing roof

Vertical ground mount

South

© CENGAGE LEARNING 2012

plus 10 to 20° is also acceptable. For instance, a location at 44° north latitude may have a slope angle of approximately 60°. Regardless of the application, all solar collectors should be mounted with a tilt angle of at least 15° so there is enough slope to ensure that normal rainfall will wash off any dirt, pollen, and so forth to help prevent soiling of the exterior glass casing. Figure 3-7 shows examples of various collector panel locations to ensure that they are facing due south and allow for the proper slope.

Mounting the Solar Panel

Most solar thermal collectors are roof mounted. In order to achieve a successful installation, several factors will need to be considered to ensure that the collector or array of collectors will maintain structural integrity with the roof and its components. In addition, the installer must take into consideration whether the collectors will be subjected to extreme weather conditions such as tornados or hurricanes. In situations such as these, the wind forces acting on an elevated or raised collector create suction on the front of the panel and an uplifting force from behind (Figure 3-8).

Figure 3-8

Wind forces acting on an elevated or raised collector can create suction on the front of the panel and an uplifting force from behind.

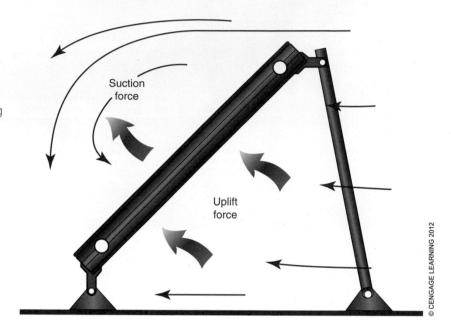

© CENGAGE LEARNING 2012

Do not tempt the strength of Mother Nature! Always anticipate the worst possible weather scenario for the area where the collectors are to be installed and plan accordingly.

The solar collector manufacturer should provide the proper framework, mounting brackets, and fasteners that will achieve a strong and stress-free installation. Before beginning the installation, verify that the roof structure is in suitable condition for mounting the collectors. An inspection should be performed that ensures the shingles are in satisfactory condition and the roof deck and truss support systems are structurally adequate to support the weight of the solar collectors. In most cases, there is usually no problem with mounting a standard collector on a conventional roof (Figure 3-9).

The first step in mounting the solar collector or array of collectors is to locate the roof trusses or rafters to which the frameworks, brackets, or clips will be mounted. This can be done by locating them in the attic, then carefully measuring the distance between them on the roof. After identifying the location of the trusses, the collector's mounting clips are attached to the trusses using lag bolts (Figure 3-10). Typically, a sealant such as silicone caulk is applied to the underside of the mounting clip to prevent water from seeping through the roof penetration.

Stainless steel fasteners are the primary choice of installers because they are less susceptible to rust and corrosion. Once the mounting clips are fastened to the roof, the next step is to install any framework that may be required. In

Figure 3-9

In most cases, solar collectors can be flush mounted on the roof.

© ISTOCKPHOTO/PAVLO VAKHRUSHEV

Figure 3-10

Detail showing lag-bolt mounting of roof bracket.

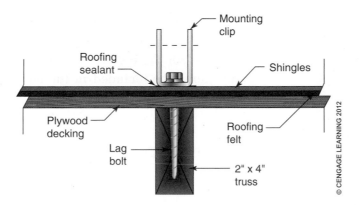

© CENGAGE LEARNING 2012

some cases, the collector is attached directly to the mounting clips. This will depend upon the pitch of the roof. When all framework is securely fastened, the collector is raised onto the roof and attached to the roof clips or to the framework. Note that flat-plate collectors can range in weight from 100 to 150 pounds, thereby requiring more than one individual to assist in mounting them in place (Figure 3-11).

Once the collectors are in place, the next step is to install the solar collector piping and make the proper roof penetrations where the piping will be run through. This process involves running the supply and return fluid piping through the proper roof flashing and sealing this flashing to ensure that the project will be leak free (Figure 3-12).

If there is any doubt in performing this operation, consult a local roofing contractor for advice or to assist in this step. Remember that typically there will be

Figure 3-11

Collectors can weigh over 100 pounds and require more than one individual to lift them into place.

temperature sensor wiring that will be run to the roof as well. This sensor measures the temperature of the water leaving the solar collector and is used to control the circulating pump. For those systems that use this wiring, a special fitting should be included in the flashing cap for the wire routing and sensor to be protected from damage.

System Piping

Once the solar collector or array has been successfully mounted, the piping is then installed below the roof to connect the collector to the storage tank (Figure 3-13). Depending upon the type of system that has been chosen (active or passive, open or closed loop), this piping arrangement may vary considerably. Consult Chapter 2 to determine the type of piping arrangement that is best suited to the particular application that has been chosen.

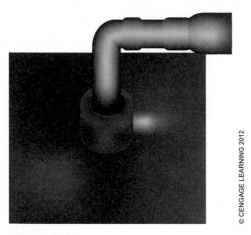

Figure 3-12

Piping must be routed through the proper roof flashing and sealed properly to ensure that it will be leak free.

Regardless of the solar thermal storage application, there are typical piping practices and accessories that should be adhered to, especially with a closed-loop/pressurized system. The typical accessories used in a closed-loop system are described next.

Expansion Tank

An expansion tank is a small pressurized vessel used to divert the expansion of the water or water/antifreeze mix when it becomes heated. When water becomes heated, it expands. If this expansion is not allowed to be diverted or relieved in

Figure 3-13

A typical piping configuration for a solar thermal storage unit.

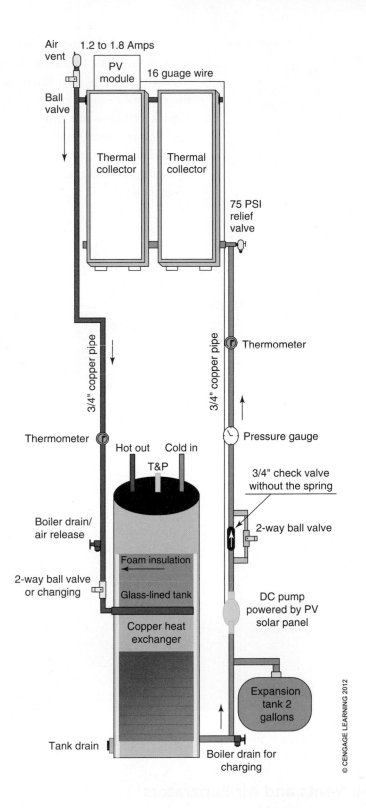

Air vent

1.2 to 1.8 Amps

PV module

16 guage wire

Ball valve

Thermal collector

Thermal collector

75 PSI relief valve

3/4" copper pipe

3/4" copper pipe

Thermometer

Thermometer

Pressure gauge

Hot out   Cold in

T&P

3/4" check valve without the spring

2-way ball valve

Boiler drain/ air release

Foam insulation

2-way ball valve or changing

Glass-lined tank

DC pump powered by PV solar panel

Copper heat exchanger

Tank drain

Boiler drain for charging

Expansion tank 2 gallons

© CENGAGE LEARNING 2012

some fashion, it could rupture the piping. The expansion tank contains a rubber bladder inside of it that separates the air from the water. When the water is heated, the rubber bladder allows the water to expand. When the water cools, the bladder contracts while pressure in the air chamber maintains a positive pressure in the closed loop (Figure 3-14).

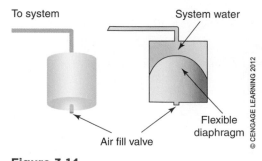

To system

System water

Air fill valve

Flexible diaphragm

© CENGAGE LEARNING 2012

Figure 3-14

A cut-away view of a common expansion tank.

This bladder must be compatible with glycol. The air side of the tank contains compressed air that can be regulated by means of a fitting used to increase or decrease the pressure—similar to filling an automotive tire with air. Typically these are pre-charged at 12 to 15 **psi** (pounds per square inch) with nitrogen. The expansion tank should also be installed downstream of the indoor tank or heat exchanger, and upstream of the circulating pump (Figure 3-15).

Figure 3-15

An installed expansion tank.

COURTESY OF DONALD STEEBY

Air Vents and Air Separators

It should be a standard practice to install an **air vent** and **air separator** on any closed-loop system. Water contains a certain percent of dissolved oxygen, and when confined to a sealed, pressurized system, this oxygen breaks free from the water and creates trapped air. If left unchecked, air in a sealed system can cause corrosion and air locks, which will prevent the water from being circulated through the system.

As the name implies, an air separator is used to separate air from the water as it flows through the system (Figure 3-16). The latest generation of air separators incorporates a mesh screen that causes the air to collide and adhere to it. As more air bubbles adhere to the mesh, they get larger, break loose, and travel up into the air vent.

Automatic air vents are usually mounted on top of air separators and contain an internal disc that swells when water comes in contact with it. This swelling seals off the air vent port. However, as air accumulates around the disc, it becomes dry and shrinks, allowing the air to pass through the vent port. As the vented air is replaced by water, the disc once again swells, closing off the port. The air vent is installed vertically at the highest point in the system, on the return line (Figure 3-17).

Figure 3-16

An air separator.

Figure 3-17

An air vent located at the highest point on the loop.

COURTESY OF DONALD STEEBY

Pressure-Relief Valves

Pressure-relief valves, sometimes referred to as safety relief valves, are used to prevent the pressure from becoming too great in the solar thermal storage system (Figure 3-18). These valves can be obtained with a fixed or adjustable pressure setpoint. When the calibrated pressure is reached, the valve opens, releasing the fluid to the atmosphere and preventing the system pressure from reaching unsafe levels that might damage the collector and related equipment. Pressure-relief valves are typically set for between 30 and 75 psi, depending on the configuration of the given system. They should be installed near the bottom header of the collector.

Freeze-Protection Valves

Freeze-protection valves (FPVs) are also called freeze-prevention valves and are obviously used in geographic areas where freezing pipes can be an issue. They are installed near the outlet of the roof-mounted solar collector to protect the collector tubes from freezing during cold weather. The freeze valve should also be installed so that it not only protects the collector from freezing, but also protects the piping between the collector and the valve. In an indirect thermosyphon system, the FPV can also be used to protect both supply and return piping. Freeze protection valves contain a wax-like material that fills a small enclosure. When it reaches the

Figure 3-18

A pressure-relief valve.

COURTESY OF DONALD STEEBY

© CENGAGE LEARNING 2012

Figure 3-19

Freeze valves are used to protect the collector tubes from freezing during cold weather.

freezing point, this material changes volume, inducing a small flow of water through the relief port (Figure 3-19).

Check Valves

A **check valve** permits fluid to flow in one direction only. It also reduces the amount of heat loss in the system at night by preventing a convectional flow of heat from the warm storage tank to the cool roof-mounted solar collectors. This is especially true if the system is using glycol as antifreeze, due to the fact that glycol will siphon faster than water when it is cold. The check valve can be installed on either the supply or return line near the indoor storage tank. It should be mounted on a vertical line where the greatest amount of fall would occur. It is recommended that a spring-type check valve be used versus a swing-type valve, which does not seat well enough to prevent thermsyphoning. The valve itself can only be opened by the force of the water or antifreeze overcoming the closing tension of the spring (Figure 3-20). When the circulating pump is de-energized, the spring automatically closes the valve, preventing the backflow of water into the storage tank.

Figure 3-20

How a check valve works.

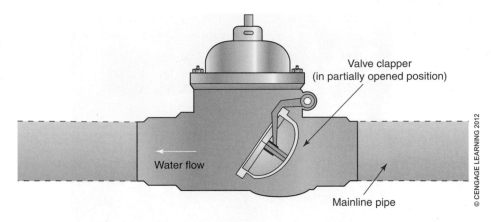

Valve clapper
(in partially opened position)

Water flow

Mainline pipe

© CENGAGE LEARNING 2012

Pressure and Temperature Gauges

Pressure gauges and thermometers are used to monitor the system's vital signs and should be installed where they can be easily read—usually at eye level. Typically, a pressure gauge with a range of 0 to 60 psi is used, especially with a glycol system. If pressure gauges are installed on each side of the circulation pump, the flow through the system can be calculated using the pump manufacturer's pump curves based on pressure drop.

A thermometer should be installed on the return side of the solar collector, before the storage tank or heat exchanger. The range of the thermometer should be 0°F to 250°F (Figure 3-21).

Figure 3-21

A combination pressure and temperature gauge.

COURTESY OF DONALD STEEBY

A second thermometer should be installed on the outlet side of the storage tank or heat exchanger. These two temperatures will allow for easy monitoring of the system's efficiency. With constant flow through the system, and under full-sun conditions, there should be a differential temperature of between 5°F and 20°F across the storage tank or heat exchanger. Two additional thermometers can be installed on the inlet and return sides of the outdoor solar collector. The differential between these two temperatures should be less than 20°F.

CONTROL STRATEGIES FOR SOLAR THERMAL SYSTEMS

Once the solar collector and indoor storage tank have been installed, and all piping has been completed, the next step is to control the system. Controls are an integral part of the total solar thermal energy system, and are vital to ensuring that the system runs at peak efficiency. No matter what type of system is being controlled, the principles of control logic are similar. Following are definitions used in control systems for solar thermal energy:

- **Controlled medium**: The substance being controlled in the system. In this case, the controlled medium is the water/antifreeze solution that flows through the loop.
- **Controlled medium temperature**: The actual temperature of the substance being controlled. In this case, the controlled medium temperature is the temperature of the water/antifreeze solution that flows through the loop.
- **Controlled device**: The device that regulates the flow through the system. In this case, the circulating pump is the controlled device.
- **Temperature setpoint**: The desired temperature of the controlled medium.
- **Inputs**: Typically, these are temperature sensors that are connected to the controller.
- **Outputs**: The signals from the controller used to energize/de-energize the circulating pump.
- **Controller**: The device that receives signals from the temperature sensors, compares them to the setpoint value, and sends the appropriate output signal to the controlled device.
- **Control loop**: The arrangement of the input device, controller, and output device within the system.

One of the most common control strategies for active solar thermal energy systems with an enclosed glycol loop is the use of differential temperature. This type of system monitors two temperature sensors as a means of controlling the respective circulating pumps. One sensor is located at the outlet of the solar collector. The other sensor monitors the temperature of the water in the storage tank (Figure 3-22).

The controller is constantly monitoring the difference in temperature between these two sensors and their respective setpoints (Figure 3-23). When the temperature of the sensor located at the collector outlet exceeds its setpoint, the controller energizes the circulating pump for the glycol loop. Likewise, when the temperature of the storage tank exceeds its setpoint, the potable water-loop circulating pump is

Figure 3-22

A type of controller that monitors two temperature sensors as a means of controlling the circulating pump.

Figure 3-23

The inside of a temperature controller.

energized. The normal differential setpoint on the controller is typically between 5°F and 10°F. As long as the temperature at the outlet of the collector or at the storage tank is 5°F to 10°F higher than its setpoint, the respective pump will run continually. This would be the normal scenario during a sunny day. When these temperatures are equal to or below setpoint, the pump will shut off. Advanced controllers can also operate the circulating pump at variable speeds that are proportional to the differential temperature between the collector and storage tank. As the temperature differential rises, the pump speed increases. This strategy is used to reduce the electrical consumption of the pump under partial-sun conditions.

As an additional feature, there are circulating pumps available today that run on **direct-current (DC)** power and are energized by photovoltaic solar cells. The strategy behind this concept is to reduce the operating cost of the circulating pump and take advantage of the sun's rays to modulate the pump speed. As the sun's energy level increases, the circulating pump automatically increases its speed, allowing for greater thermal transfer through the collector (Figure 3-24).

Another strategy is to use both a pump powered by **alternating current (AC)** and a DC-powered pump, so that if power is lost to the AC pump, the DC pump can maintain flow through the system. When incorporating DC pumps with photovoltaic modules, be sure to match the proper pump with its appropriate solar panel. Also, be sure that the DC pump has enough capacity to circulate fluid through the solar loop. Not all DC pumps have the same rated output as AC pumps, and antifreeze is more difficult to pump than straight water due to its higher viscosity.

Today's modern digital controllers incorporate the use of specialized temperature sensors called thermistors for controlling solar thermal systems. By definition, a

Figure 3-24

A circulating pump that runs on AC power and is energized by photovoltaic solar cells.

COURTESY OF DONALD STEEBY

Figure 3-25

A strap-on-type temperature sensor.

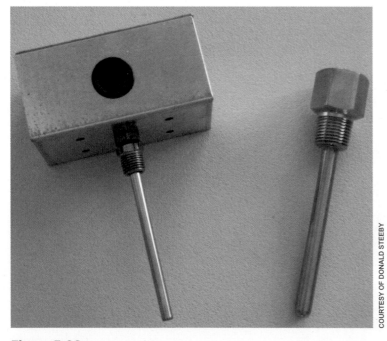

COURTESY OF DONALD STEEBY

Figure 3-26

A well-type temperature sensor.

thermistor is a resistor made of a semi-conductor material in which the electrical resistance varies with a change in temperature. As the temperature of the thermistor increases, the ohms resistance decreases. Thermistors are utilized in modern control sequences because of their accuracy and reliability. They can be installed by two different methods. One way is to attach the sensor directly to the piping to measure the water temperature. This is commonly known as a strap-on temperature device, and the installer should make sure the sensor is making good contact with the pipe (Figure 3-25).

The other method is to install a recessed well into the piping and insert the sensor into the well. This method can be more accurate than the strap-on sensor and is less susceptible to damage and neglect (Figure 3-26).

Other features that may be included with today's modern digital controllers include:

- Controlling multiple-stage units
- Liquid crystal displays (LCDs) on the controller
- Demand limit strategies
- Time clock scheduling
- Outdoor temperature reset function
- Night setback function
- Data logging

FILLING AND STARTING UP THE SYSTEM

When the temperature controls have been installed and all piping is complete, it is time to fill the system and perform a proper start-up procedure. However, before the system is filled, it must first be properly leak tested. It should be noted at this point that the expansion tank and air vent can be installed after the system has been leak tested and cleaned. Temporarily cap off the fittings where these components are to be installed.

Cleaning and Flushing the System

First fill the system with air until the pressure is between 50 and 60 psi. It should be noted that leaks will be easier to detect with air than with water. Once the system is pressurized, use a solution of mild soap and water to swab every joint and fitting for potential leaks. Dishwashing soap is preferable, or a commercial leak-detecting fluid is available at any HVAC supply house. Record the pressure on the pressure gauge and wait for at least 30 minutes to see if it falls. If possible, keep the system pressurized overnight before filling to ensure that it is leak free.

Once the system has been verified as being leak free, the next step is to clean the lines. Use a mixture of 1 cup of trisodium phosphate (TSP) per 1 gallon of water for a cleaning solution. It will probably require several gallons of solution to ensure the system gets cleaned properly. A positive displacement pump works best for pumping the solution into the system. Once filled, run the circulating pump for 30 minutes to clean out any flux, pipe thread compound, or other debris from the lines. Then drain the cleaning solution and flush the system with clean water.

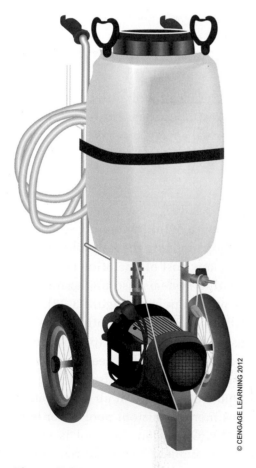

© CENGAGE LEARNING 2012

Figure 3-27

A flush-and-fill system pump.

Charging the System

At this point, the expansion tank and air vent should be installed. However, before installing the expansion tank, be sure to charge it with air to the correct pressure, typically 15 psi. It will not harm the system if the tank exceeds 15 psi, as long as this pressure does not exceed the manufacturer's recommendations for the solar collector or system components.

Before filling the system with water or glycol solution, cover the solar collector to prevent the solution from being heated too quickly while filling. Now open the air vent located near the solar collector. This will force the system's air to the highest point of the loop. The pressure-relief valve may also be opened for the initial fill. Close it once the fluid begins to flow out of it. Once again, a positive displacement pump is best used for charging the system. Introduce the fluid solution at the lowest point of the system to force the air upward to be displaced into the atmosphere. A flush-and-fill cart can come in handy at this point (Figure 3-27).

Once the system has been filled, energize the circulating pump to force any air bubbles out of the piping. The air vent should remain opened until all air bubbles have been displaced from the system. Record the operating system pressure and monitor this pressure for several days to ensure it stays constant. If during this period a pressure drop of less than 10 psi is observed, then more than likely the system is air free.

The temperature controls can now take over the monitoring and control of the system. Over the course of the first several weeks of operation, periodically check the system pressures and temperatures to ensure they are in range. As mentioned earlier, the temperature through the heat exchanger or storage tank should maintain a differential of between 10°F and 25°F. This will indicate that the heat exchanger is functioning properly.

Additional maintenance points to monitor or observe are as follows:

- Take note if the pump is running at the "wrong time," such as at night.
- Wash the solar collector annually with water and a soft brush.
- Spray the collector with water during prolonged periods of no rain.
- Install labeling on all piping showing the direction of flow.
- Periodically check all fittings for possible leaks.
- Oil the circulating pump motor on a semiannual basis if required by the manufacturer.
- Unless there is a problem with stagnation, the glycol solution should not need to be replaced for 10 years.

VARIOUS APPLICATIONS FOR SOLAR THERMAL SYSTEMS

At this point, most of the information presented has dealt with utilizing the solar thermal storage system for domestic hot water applications. However, several other applications are just as suited for solar thermal energy systems.

Swimming Pools

The intent of any pool heater is to extend the length of the swimming season by beginning earlier in the spring and extending this time into the fall. What makes solar pool heaters so attractive is the relatively low cost of operation compared to conventional pool heaters. Solar pool heating systems use unglazed, low-temperature collectors typically made from polypropylene (Figure 3-28). The intent of using this type of collector is for it to operate just slightly above the temperature of the surrounding air. By doing so, the collector is capable of heating a large volume of water only a few degrees. Remember that heating a swimming pool with a solar collector is a marathon, not a sprint. Also, raising the temperature of the water only a few degrees above the ambient air temperature will prevent heat stratification and maintain a more consistent pool water temperature. By implementing a pool cover or blanket when the pool is not in use, water temperatures of 18°F to 25°F above ambient temperatures can be achieved.

The pool's solar collector is typically sized to be 50% to 100% of the surface area of the pool. A rule of thumb is that a 3°F temperature rise can be achieved for each 20% of the pool surface area that is added to the solar collector. For instance, an

Figure 3-28

A typical piping arrangement for a solar pool heating system.

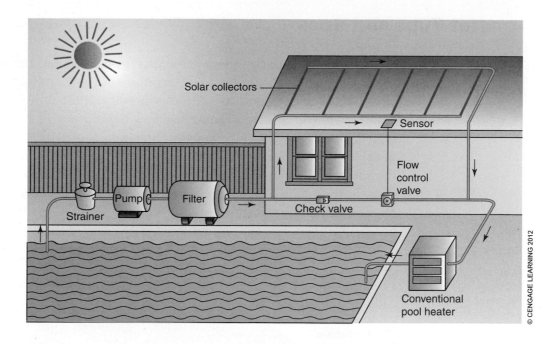

8°F temperature rise can be achieved for a pool area of 512 sq. ft. and a collector size of 256 sq. ft. However, the temperature rise will increase to 15°F if the collector size is increased to 512 sq. ft. for the same size pool.

Pool solar collectors are best installed on south-facing roofs, although any location that is unshaded during the middle 6 hours of the day will work, even for a free-standing collector. In fact, 80% of the solar radiation gained from the collector will typically occur during a 4-hour period. For a south-facing solar collector, this would be between the hours of 10:00 a.m, and 2:00 p.m. Remember that shading from trees or nearby buildings will greatly reduce the capacity of any solar pool collector.

Another added benefit of using solar energy to heat swimming pools is that the existing circulating pump can be used to circulate water through the collector as well as through the pool filtration system. Just as with the domestic water heating application, the solar pool heater incorporates two temperature sensors for controlling the system. One sensor is located in the PVC piping downstream of the circulating pump. It measures the actual temperature of the water leaving the pool. The other sensor is located in the middle of the collector array. The difference with the solar pool heating system is that it uses a diverting valve. When the controller senses that the collector temperature is 5°F to 8°F warmer than the pool water sensor, it energizes the diverting valve to direct the flow of water to the collector, and then back to the pool. When the controller senses that the differential temperature between the collector and the pool is too low, it de-energizes the diverting valve, which blocks the flow of water to the collector. The pool water sensor also acts as a high limit that stops the flow to the collector, should the pool water temperature exceed its setpoint.

Solar pool heating systems generally cost between $7 and $12 per square foot of pool area, depending upon system design and collector type. This can provide a return on investment of between 1.5 to 7 years as compared to conventional heating methods. Clearly, solar pool heating is one of the most cost-effective uses of solar energy on the earth.

Hot Tubs and Spas

Solar heating of hot tubs and spas is similar to pool heating (Figure 3-29). However, in order to achieve the increased temperatures of 100°F and higher, a dedicated solar collector may be necessary. Also, because it is difficult to achieve these higher temperatures during the evening hours, a backup source of heat is usually required. One strategy for incorporating a hot tub along with an existing swimming pool is to use the pool's collector to increase the tub or spa temperature along with the pool's water temperature, and then trim out the spa temperature using an auxiliary heating source. As with swimming pools, the use of a spa cover is essential for saving money on energy costs.

Figure 3-29

A piping layout for a solar hot tub.

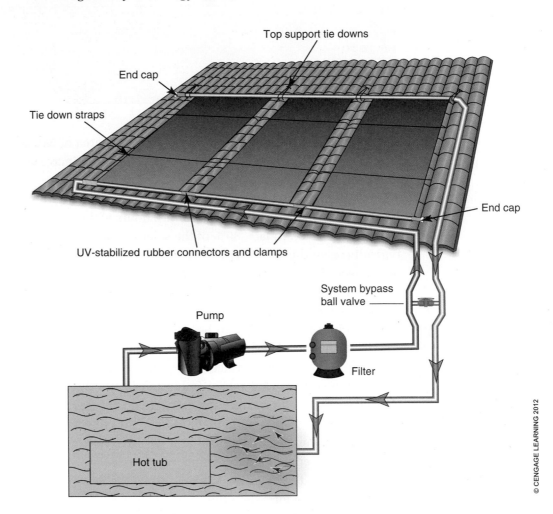

Space Heating

The two main methods of utilizing solar energy for space heating are classified as radiant heating and forced-air systems. Both types of systems are dependent upon whether the structure is being considered as new construction or is being retrofitted for space heating needs. Generally, it is a better idea to incorporate a solar thermal space heating system into a well-insulated, tightly constructed building.

Therefore, this type of system works well for new construction. Regardless of which type of system is preferred, utilizing solar thermal energy systems for space heating will usually satisfy 50% to 80% of the annual space heating loads.

Radiant Solar Heating

Radiant heating is accomplished by circulating hot water through a network of tubing, usually embedded in the floor, or through terminal heating devices such as convectors or radiators. Once the device is heated, it radiates warmth throughout the conditioned space. The defining characteristic of solar radiant heating is the low water temperature at which it operates. Whereas conventional radiant heating systems operate at water temperatures of up to 180°F, solar radiant typically performs at a water temperature of no greater than 120°F. The reason being is that lower design temperatures allow for greater solar energy yields, and these lower temperatures equate to a reduction in the size of the solar collector. Radiant solar heating is typically delivered to the space either through heated floor slabs or through radiant panels (Figure 3-30).

Heated floor slabs incorporate plastic or **cross-linked polyethylene (PEX)** tubing that is spaced close together and embedded into the subflooring of the structure. This design is well suited for use with floor covers that have a low **R-value,** such as tile, vinyl, or hardwood surfaces. The use of radiant floor heating with carpeting

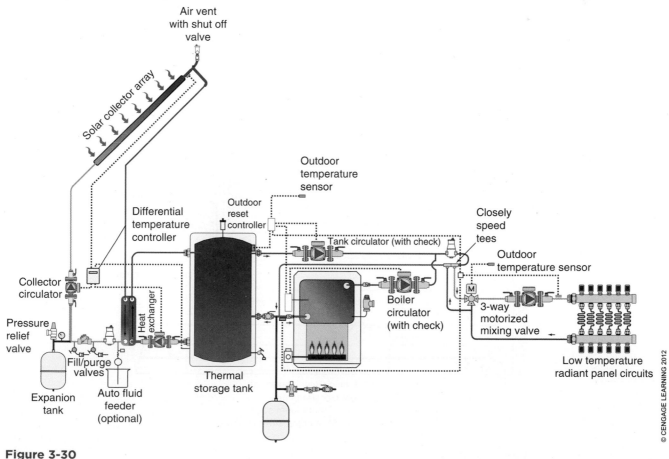

Figure 3-30

A piping arrangement for radiant floor heating.

is not recommended, as the carpet will act as an insulated barrier between the heated surface and the occupants. The tubing is typically spaced 6 inches apart to maintain a room temperature of 70°F if the water in the circuit can maintain at least 88°F.

Spacing of the tubing can be increased up to 12 inches apart if the supply water temperature can maintain a range of between 95°F and 98°F. Tube sizing for radiant floor heating is typically ½″ in diameter. Separate zones can be controlled independent of each other by means of a space thermostat controlling a zone valve. The main water supply line is connected to a manifold station, where individual zones are circuited to and from. Each circuit contains a zone valve that opens when an individual thermostat calls for heat.

Radiant panels can be integrated into walls and ceilings and offer an alternative to floor heating (Figure 3-31). Just as with solar floor heating systems, radiant panels operate at relatively low supply water temperatures. To achieve an adequate heating output, these panels require a high surface area relative to the rate of the heat to be delivered to the space. They also favor a relatively low internal resistance between the tubing and the surface area releasing heat into the space. Radiant panels are typically covered with drywall or a similar wall covering, which make them indistinguishable from a standard interior wall or ceiling. Because they typically have a low thermal mass, radiant panels respond quickly to changes in internal load conditions or zone setback schedules. With a supply water temperature of 110°F, the typical radiant panel will deliver approximately 32 BTU/hr/ft² in order to maintain a space temperature of 70°F. A heating load analysis on the conditioned space will determine the required BTUs to heat the space. Once this is known, the correctly sized panel can be incorporated. Most manufacturers can provide output ratings for their respective radiant panels.

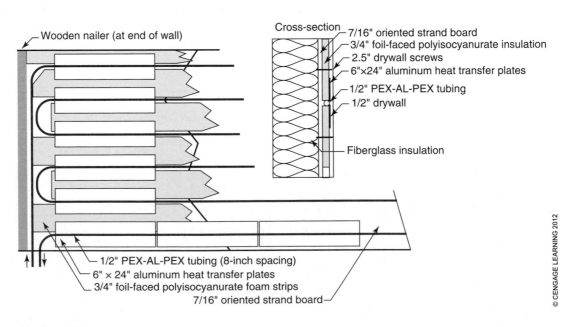

Figure 3-31

A layout for radiant wall heating.

Forced-Air Systems

The most common application for using solar energy with a forced-air system involves the use of a water/glycol mix that is circulated through a hot water coil. The coil is typically installed in the discharge plenum of a forced-air furnace. Care must be taken to ensure that the coil is placed downstream of the furnace's heat exchanger and air conditioning coil. Failure to do so could potentially void the warranty on these components. In addition to the hot water coil, a three-way mixing valve and supply air temperature sensor are installed for control purposes. The supply air sensor is installed downstream of the coil and is used to regulate the discharge temperature from the furnace. On a call for heat, the furnace blower is energized, and the mixing valve modulates the flow of hot water to the coil based upon the supply air temperature. When the solar energy system cannot keep up, the furnace burner is energized to maintain proper space temperature (Figure 3-32).

As an additional measure, a storage tank is incorporated into the solar thermal storage system to ensure enough hot water is available to the coil. A separate control system is used to maintain the hot water temperature in the storage tank, similar to a domestic hot water application. A check valve is installed in the piping that supplies water to the heating coil to prevent reverse heat migration from the furnace to the solar storage tank when the coil is not being used.

The best economical solar energy system sizing will usually satisfy 50% to 80% of the annual space heating loads. Maximum efficiency is achieved if the solar space heating system is operated during prime solar time, which is typically from 9:00 a.m. to 3:00 p.m. This scenario makes solar forced-air heating an excellent application for commercial businesses that operate during this time frame.

The amount of solar collector area needed to heat with radiation or with forced air depends upon many factors. These include the amount of solar energy that is available, the solar collector's efficiency, local geographic climate, and the heating

Figure 3-32

A typical piping arrangement for a solar space heating application.

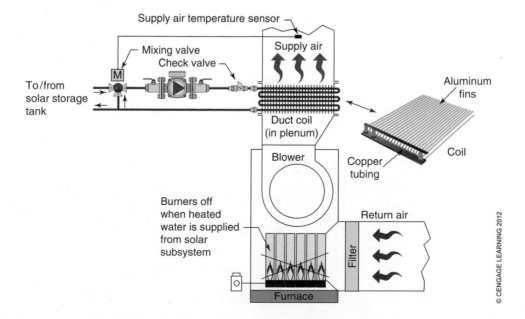

requirements of the conditioned space. Heating requirements are based on insulation levels, the tightness of the house, and the lifestyle of the occupants. Generally, the area of solar collector needed is approximately equal to 10% to 30% of the building's floor area in square feet.

SOLAR COOLING APPLICATIONS

At first it would appear that the term "solar cooling" would be a bit contradictory. After all, harnessing heat from the sun is the main purpose behind using solar thermal energy, not cooling. However, there are practical applications for using the sun's energy to transfer heat out of a building as well as adding heat to it. One of the simplest ways of taking advantage of solar cooling is through the use of a solar photovoltaic panel to power the outdoor fan motor on a condensing unit. During peak daylight hours this application can be a money-saving alternative. However, solar energy can best be utilized with air conditioning when it is used in conjunction with **absorption cooling** (Figure 3-33).

The absorption cooling process uses a source of heat to provide the energy needed to drive the cooling system, rather than a mechanical source, such as what is used on a conventional air conditioner. In this case, the source of heat is water circulating through the solar collector. In some cases the amount of heat needed to drive the absorption unit can be as much as 190°F. For this reason, a vacuum-tube solar collector may be required to generate enough hot water to run the system effectively.

Instead of using a conventional refrigerant, such as R-22, the absorption system typically uses ammonia and water as a medium for heat transfer. Just as with conventional air conditioning systems, the absorption unit incorporates an **evaporator** and **condenser.** Heat is absorbed into the water as it passes through the evaporator and is then rejected in the condenser. The water acts as a refrigerant as it condenses from a vapor to a liquid under high pressure in the

Figure 3-33

Using solar energy to power an absorber.

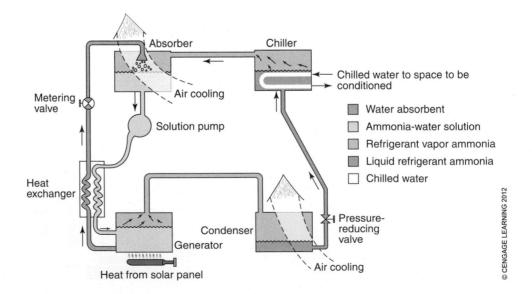

© CENGAGE LEARNING 2012

condenser. This high-pressure liquid passes through a restrictor, where the pressure and temperature are reduced. The low-pressure liquid boils off inside the evaporator, absorbing heat in the process. From there, the vaporized water travels to the absorber, where it is mixed with ammonia. The ammonia has an affinity for the water and as it is absorbed into the ammonia, part of it condenses back into a liquid. This is where a great deal of heat transfer occurs. However, the absorber cannot cool the solution adequately enough to remove all of the heat required; therefore much of the water remains in a vapor state within the ammonia solution. To remove the required additional heat, air is forced over the water/ammonia solution, which allows the additional heat to be extracted and the water vapor to be condensed. A solution pump then forces the water/ammonia mix into the generator. This is where the solar collector comes into play. The collector heats a separate water supply that is used to transfer heat to the generator. The generator is where the water is boiled out of the ammonia until it vaporizes and is sent back to the condenser, where the process begins all over again.

Absorption cooling is not a new technology; in fact, it has been around for over 150 years. Any given recreational vehicle may be using an absorption refrigerator powered by propane gas. Another heat-transfer solution that is commonly used for commercial absorption chillers instead of ammonia is **lithium bromide.** This solution, however, may not be practical for residential or light commercial applications due to the fact that lithium bromide requires a much higher generator temperature for it to function properly. Most solar thermal energy systems cannot generate the level of heat that is required. The use of absorption cooling that incorporates solar thermal energy is currently limited to a residential or light commercial basis. This is due to the fact that equipment costs are high and there is still debate as to whether solar collectors can generate enough heat to make the system economical. However, technology is constantly advancing, and as it does, equipment costs should decrease while efficiency levels increase.

Case Study

—*Dave and Sharon Kaechele*

Still Working After All These Years

The 1970s was a thriving era for solar installations. Because of the increasing costs of fossil fuels and because of attractive government incentives on new solar equipment, many homeowners took advantage of the fact that they could save money on energy and at the same time reduce their costs for equipment and installations. Dave and Sharon Kaechele were no exception. They purchased a hybrid package in 1977 that consisted of a solar thermal storage system and a wood stove (Figures 3-34 and 3-35).

Figure 3-34

Roof-mounted solar collector.

COURTESY OF DONALD STEEBY

Figure 3-35

Wood stove used for water heating and domestic comfort heating.

© ISTOCKPHOTO/PIXELBRAT IMAGERY

Case Study (Continued)

Their total cost for the package was $2,200. However, they received a 50% return in government tax rebates, making the system price look very attractive.

Their solar thermal storage system consists of a glycol-filled closed loop and a plate-type heat exchanger.

The other side of the heat exchanger heats the domestic water supply for their house. In the summertime, the solar storage system takes care of all of their hot water needs. In winter, the wood stove supplements the solar panel when there is not enough sunshine. At peak solar hours, the domestic water can reach 120°F, and Dave has the system plumbed so that he can alternate between solar and wood heat or use a combination of the two (Figure 3-36).

Figure 3-36

Hot water storage tanks.

COURTESY OF DONALD STEEBY

There is an electric heating element in their water heater, but it is seldom needed. The wood stove is also their primary source of whole-house heating in the winter.

The only real maintenance that Dave performs is to check the operating pressures in the hot water loop annually and occasionally bleed off some excess air. He has replaced the antifreeze in the system just once since the day it was installed.

After 30+ years of trouble-free operation, the system has paid for itself many times over. Dave and Sharon are living proof that alternative energy systems can withstand the test of time, and still keep working after all these years.

The science of solar photovoltaics (PVs) is the process of converting solar energy from the sun into electrical energy. The electrical energy that is generated from sunlight is used for many different purposes, such as pumping water, lighting up the night, charging batteries, running motors, supplying power to the electrical grid, and much more (Figure 4-1).

At any given location during a clear day, the amount of sunlight that is striking the earth is equivalent to approximately 1,000 watts of power per square meter.

Science has been researching photovoltaics since 1839, when French physicist Edmund Bequerel found that certain materials would produce small amounts of electrical current when exposed to light. In 1873, British scientist Willoughby Smith discovered that **selenium** was also sensitive to light. He also discovered that the conductivity of selenium increased in direct proportion to its exposure to light. As a result of Smith's work, Charles Fritts developed the first selenium-based solar cell in 1880. This cell produced electricity without the consumption of any materials, and without the generation of any heat. It wasn't until 1905, when Albert Einstein described the nature of light and the photoelectric effect, that photovoltaics finally received broad acceptance as a legitimate source of power. Later on, Einstein won a Nobel Prize in physics for his work in this area.

The first photovoltaic module was built by Bell Laboratories in 1954 as a means of generating dependable power to remote communication systems. These scientists discovered that **silicon** was also sensitive to light, and if treated with impurities, would generate substantial voltage. This module was called a solar battery and could achieve an efficiency of only 6%. It was not until the 1960s when NASA installed a PV system on its first satellite that photovoltaics was considered a serious technology (Figure 4-2).

Figure 4-1

The electrical energy that is generated from sunlight is used for many different purposes.

Figure 4-2

NASA installed a PV system on its first satellite in the 1960s.

As a result of the space program, PV technology had finally advanced to where it is today. In addition to this, the energy crisis of the 1970s gave photovoltaics the recognition it deserved as a legitimate source of power for non-space applications. Today, solar modules supply electrical power to over 1 million homes around the world. The current market for solar photovoltaic systems is growing at approximately 17% per year, while the cost for a system is about 45% lower than it was just 10 years ago. Meanwhile, the life expectancy for a current solar PV module is almost 30 years.

SOLAR PHOTOVOLTAICS ON AN ATOMIC LEVEL

Most solar cells are made from silicon, the 14th element on the periodic table and the second most abundant element found on earth. Silicon is also the same semiconductor material that is used in the microelectronics and computer industries. As a semiconductor, silicon has the properties of being both a conductor and an insulator of electricity. The electrical conductivity of silicon can be varied by introducing small amounts of impurities, called **doping**, into the semiconductor structure. The impurities that are used are typically boron and phosphorous. These elements create a permanent imbalance in their molecular charge, which enhances the silicon's ability to distribute an electrical charge.

The solar cell (Figure 4-3) is actually a thin semiconductor wafer of silicon that has been doped to form an electric field—positively charged on one side with boron (**p-type material**), and negatively charged on the other side with phosphorous (**n-type material**).

When light, which is composed of energy particles called photons, strikes the solar cell, it knocks loosely held electrons from the negative side of the wafer. These excited electrons are attracted to the positive side of the solar cell. If electrical conductors are attached to the positive and negative sides of the solar cell, an electrically charged circuit is created that can be used to power a load, such as a light bulb. The amount of **electromotive force** created by one solar cell is approximately 1/2 volt. This voltage does not vary according to the size of the

Figure 4-3

The anatomy of a solar cell.

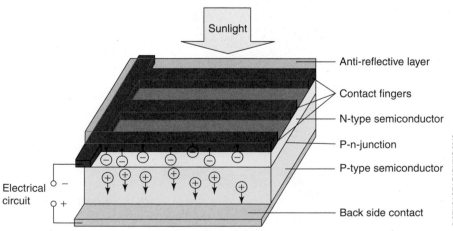

Figure 4-4

The graduation of a solar cell to a solar array.

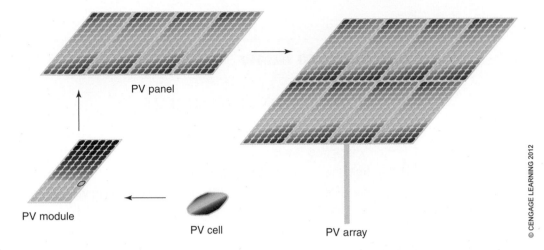

PV panel

PV module

PV cell

PV array

© CENGAGE LEARNING 2012

solar cell; however, the amount of current will vary. By following this principle, the larger the area of the cell, the greater the amount of current that can be produced.

Once the solar cell has been produced, it is covered with an antireflective coating to enhance the amount of sunlight that it can absorb. The individual cells are then wired together to achieve the desired voltage and current to form a PV module. The solar cells are also encapsulated within the module's framework to protect them from inclement weather. Each module contains approximately 36 to 40 solar cells. Because each cell produces approximately 1/2 volt, a 36-cell module will typically produce an operating voltage of about 18 volts under standard test conditions (STCs). The nominal voltage of this module would be 12 volts. Typically, **photovoltaic panels** are flat, rectangular panels that can produce between 5 and 300 watts of electrical power. A solar panel is generally one or more modules that have been wired together. An array is a group of solar panels that can be connected in both series and parallel electrical arrangements to produce any required voltage and current combination for a given application (Figure 4-4). Solar arrays are usually roof mounted, but can be fastened to a bracket and frame for ground mounting as well (Figure 4-5).

© ISTOCKPHOTO/QUANG HO

Figure 4-5

Solar arrays can be pole mounted as well as roof mounted.

Things to Know

USING SILICON IN SOLAR CELLS

One reason that silicon is used to create solar cells is that the energy needed to "knock free" a silicon electron matches the energy created by photons coming from the sun. If the sun's photons had less energy, they would not be able to free the silicon electrons, and if they had more than what was necessary to free the electrons, the energy would be lost as heat and wasted.

PHOTOVOLTAICS AND ELECTRIC PRINCIPLES

In order to understand the workings of solar photovoltaics and their applications, there must be a thorough understanding of the principles of electricity. Electricity in its simplest form is the flow of electrons through a circuit. There are three key components to electricity:

- Voltage
- Amperage
- Resistance

Electrical Terminology

A volt is the unit of force that causes the electrons to flow through a circuit. In most cases, the circuit is a wire. Volts are represented by the abbreviation V, and are expressed by the symbol E. Electrical pressure is also referred to as electromotive force, or EMF. To better understand the principles of electricity, an analogy can be made with water flowing through a garden hose. In this instance, voltage would represent the water pressure flowing through the hose. For most residential and commercial applications, electricity is delivered at 120 and 240 supply voltages to the structure from the utility company.

Amperage is expressed in amperes or amps and is the unit of electrical current flowing through the circuit or wire. Amps use the abbreviation A and are expressed by the symbol I, which stands for the intensity of the current. All wiring used in electrical circuits is sized according to the amount of electrical current or amperage flowing through it. Using the garden hose again as an example, current would be considered the volume of water that is flowing through the hose in gallons per minute.

Resistance is an internal property of matter that resists the flow of electrical current. Resistance is expressed by the abbreviation R, and by the symbol Ω (omega). Electrical conductors such as copper wire have a very low resistance. However, insulators such as rubber or glass have a very high resistance. The resistance of an electrical circuit is measured in ohms, which is named after the German physicist Georg Ohm. With reference to the garden hose analogy, resistance would be considered the length of the hose. A longer hose would equate to a greater drop in water pressure. Similarly, a longer electrical circuit will have a voltage drop as a result of an increase in the resistance of the length of wire. The resistance through a given electrical circuit is dependent upon three factors: the length of the wire, the cross-sectional area of the wire, and the quality of its material. A larger-diameter wire can carry more current due to the fact that it has a lower resistance, and as a result, a higher number of electrons can pass through it at one time.

The relationship between voltage, amperage, and resistance is defined by **Ohm's Law**. This law states that the current passing through a conductor is directly proportional to the voltage, and inversely proportional to the resistance (Figures 4-6 and 4-7). When using the following symbols:

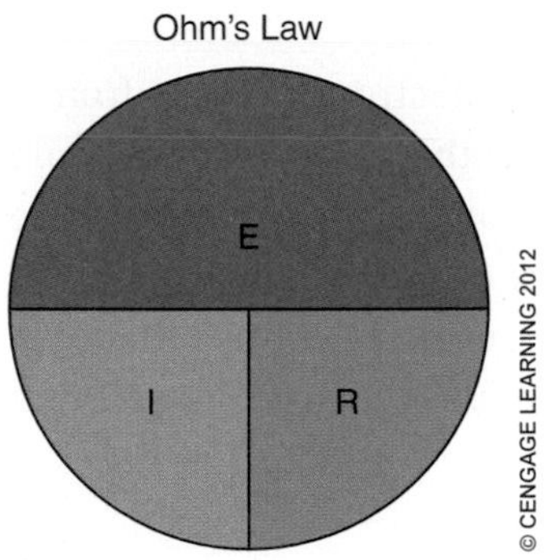

$$E = \text{Voltage}$$
$$I = \text{Amperage}$$
$$R = \text{Resistance}$$

Ohm's Law can be written as:

$$E = I \times R$$
$$I = E/R$$
$$R = E/I$$

Ohm's Law

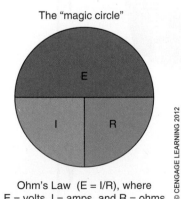

© CENGAGE LEARNING 2012

Figure 4-6

This symbol shows the relationship between volts, amps, and ohms, as used in Ohm's Law.

The "magic circle"

Ohm's Law (E = I/R), where E = volts, I = amps, and R = ohms

Figure 4-7

To determine the formula for an unknown quantity using Ohm's Law, cover the letter that represents the unknown.

There is a fourth unit of measurement that is used in electrical principles. The watt is a unit of power that indicates the rate at which a given load is consuming electrical energy. Watts are calculated by multiplying the volts time the amps within an electrical circuit. In order for an electrical utility to determine the amount of electricity used by a consumer, kilowatt-hours are measured for billing purposes. A kilowatt is equal to 1,000 watts of electrical power, and a kilowatt-hour is a measurement of power consumed over time.

Types of Electrical Current

There are two types of current found in electrical circuits: **AC**, or **alternating current**, and **DC**, otherwise known as **direct current**. Alternating current is electrical current that flows in reverse or alternating directions at frequent and regular intervals. It is produced by an electrical generator or alternator, which determines the **frequency** in hertz (Hz). Frequency is measured in cycles per second; the normal frequency for electricity in the United States is 60 Hz. Because the voltage can be readily changed through the use of transformers, AC electricity is more suitable for long-distance transmission. AC can also employ capacitors and inductors, allowing for a wide range of electrical applications.

Direct current is electrical current that flows in only one direction. It is the continuous movement of electrons from a negatively (−) charged area to a positively (+) charged area through a conductor. An example of direct current is the positive and negative terminals of a battery. These terminals are always considered positive and negative and the current always flows in the same direction between the two terminals.

It is important to understand that photovoltaic modules produce only direct current and that this current can be stored in batteries. In order to convert the direct current produced by the PV module to alternating current, the use of an inverter must be incorporated into the electrical circuitry.

PHOTOVOLTAIC ELECTRICAL CIRCUITS

By definition, an electrical circuit is a continuous path of electrons flowing from a voltage source through a conductor, to a load, then back to the source. An example of the voltage source would be a battery, the conductor would be the wire, and the load could be an electric motor or a light bulb. Usually a switch is incorporated into the electrical circuit to interrupt the flow of electricity and to control the load. When the switch is off, the circuit is said to be open. When the switch is turned on, the circuit is closed (Figure 4-8).

Figure 4-8

A simple DC electrical circuit.

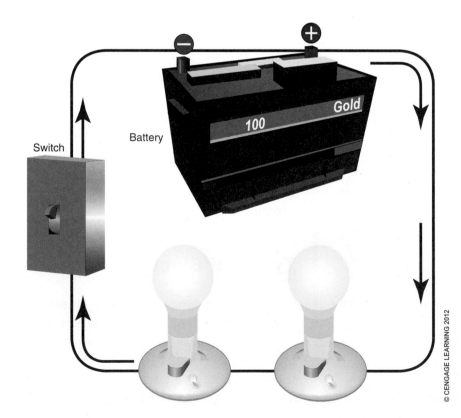

Electrical switches can be configured in many different ways. Relays and contactors are considered switches, and switches can be controlled manually, or by temperature, pressure, or time.

Series and Parallel Circuits for Power Supplies

There are three common types of wiring configurations used with PV circuits. These configurations can be used to wire either solar PV modules or the batteries used for energy storage. The three types of configurations are:

- Series circuits
- Parallel circuits
- Series-parallel circuits

A **series circuit** for a power supply such as a PV module is wired so that the positive terminal of one module is connected to the negative terminal of the next module (Figure 4-9). When power sources are wired in this fashion, the voltage supplied to the load increases. However, the amperage supplied to the load does not change. For instance, if two 12-volt modules or batteries were wired in series, a total of 24 volts would be produced. If the amperage rating of each module or battery was 3 amps, then the value of the total circuit would be 24 volts at 3 amps.

Remember that the voltage in a series circuit is additive, but the current remains the same.

A **parallel circuit** is wired so that the positive terminal of the first device is wired to the positive terminal of the next device. Similarly, the negative terminal of the first device is wired to the negative terminal of the next device (Figure 4-10).

The difference with a parallel circuit is that the resulting voltage and current output are exactly opposite of the series circuit. In a parallel circuit, the current is additive, but the voltage remains the same. As an example, if two 12-volt, 3-amp solar modules are wired in parallel, the resulting output to the load would be a 12-volt, 6-amp circuit. The same results would occur if batteries were used in lieu of solar modules. Remember that the current in a parallel circuit is additive, but the voltage remains the same.

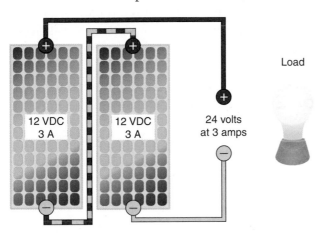

Figure 4-9

A power supply wired in series.

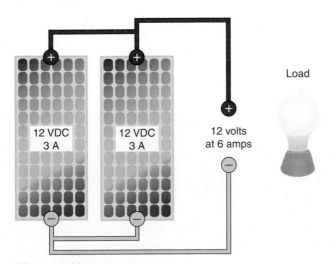

Figure 4-10

A power supply wired in parallel.

© CENGAGE LEARNING 2012

Figure 4-11

A power supply wired in a series-parallel circuit.

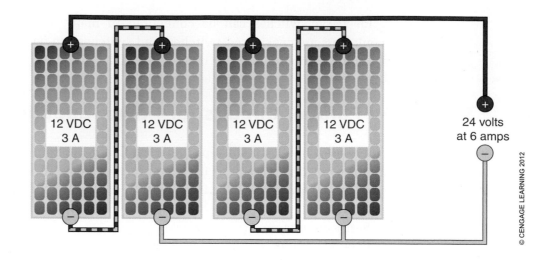

12 VDC 3 A   12 VDC 3 A   12 VDC 3 A   12 VDC 3 A

24 volts at 6 amps

© CENGAGE LEARNING 2012

Solar PV systems may also use a combination of series and parallel circuits to obtain the required voltage and amperage for a given load. These configurations are known as **series-parallel circuits.** This type of circuit is simply two or more series circuits that are wired together in a parallel configuration (Figure 4-11). For instance, if a given load or output required 24 volts at 6 amps, then four 12-volt, 3-amp solar modules could be wired in a series-parallel configuration to satisfy this demand.

Series and Parallel Circuits for Electrical Loads

Field Tip

Wiring Loads in Series

Wiring loads in a series configuration is not recommended because of two reasons: (1) Some equipment, such as electric motors, could suffer damage because of the voltage drop across the load. (2) If one of the loads fails (such as a light bulb burning out), the circuit will open and all other loads on the circuit will lose power (Figure 4-12).

Up to this point, the discussion has been on various wiring configurations for power sources. However, electric loads may also be wired in various configurations as well. When wired in either series, parallel, or series-parallel configurations, loads will react the same as power sources.

Loads that are wired in series will have a voltage drop across each load. The total drop is equal to the sum of all loads within the circuit. Conversely, the current is equal across all loads within the circuit. The following points summarize the characteristics of a series circuit:

- The electrical current can follow only one path.
- The current is the same at all points in the circuit.
- The total circuit resistance is the sum of all resistances in the circuit.
- The voltage is divided across all circuit loads.
- Any interruption in the circuit will stop current flow through the entire circuit.

Loads are typically wired in a parallel configuration, primarily because the voltage drop for each load remains equal to the source voltage (Figure 4-13). The current draw, however, will increase when additional loads are added to the circuit. Because of this, the wiring

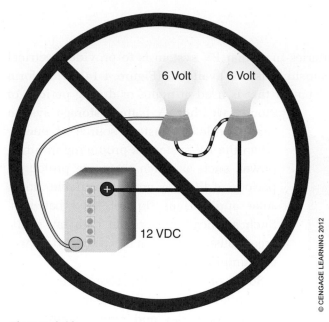

Figure 4-12

Loads wired in series. This practice is not recommended.

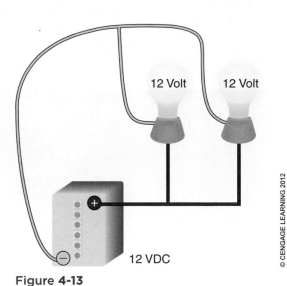

Figure 4-13

Loads wired in parallel.

size and fuse protection need to be addressed as additional loads are added to an existing circuit.

The following two points outline why practically all electrical circuits have their loads wired in a parallel configuration:

1. In a parallel configuration, each load can be controlled individually.
2. If more loads are added, the operating voltage is not affected.

The following points summarize the characteristics of parallel circuits:

- The current can take more than one path.
- Each branch circuit is unaffected by other branches.
- The supply voltage is the same in all branches.
- The current is divided between the branch circuits.
- The total circuit resistance drops as more branch circuits are added.

PHOTOVOLTAIC COMPONENTS

Once there is a clear understanding of how a solar photovoltaic module functions and how it interacts with its respective electrical circuits, it is important to understand how the other peripheral devices interact with the photovoltaic system in order to make the system fully functional. These devices include:

- Batteries
- Controllers
- Inverters

PV Batteries

The purpose of including batteries in a solar PV system is to provide electrical energy for use when direct sunshine is not available (Figure 4-14). This form of energy is considered stored power that is used at night, or during periods of cloudy weather. A battery storage system can provide a constant source of energy when the PV system is producing minimal power, such as when there are prolonged periods of little or no sun available. Some solar applications do not need batteries. Such applications include highway signs, greenhouse ventilation fans, and certain water pumping systems.

Some of these applications are specifically used during daylight hours; others can be used both day and night. Photovoltaic systems that are grid-connected to an electrical utility do not always need batteries either, although in certain situations, batteries can become useful as a source of emergency power.

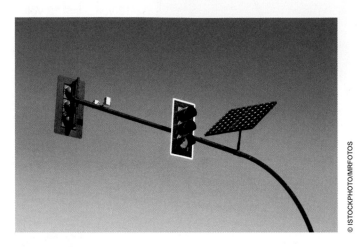

Figure 4-14

Even traffic lights can be powered with solar energy.

© ISTOCKPHOTO/MRFOTOS

The types of batteries used in solar photovoltaic applications are not typical to most applications. Most batteries, such as automotive types, are used only to produce a large current output for a short period of time, and then they are re-charged by the alternator over time, giving them a long lifespan. Solar batteries are considered deep-cycle batteries and produce a much smaller current output over a long period of time.

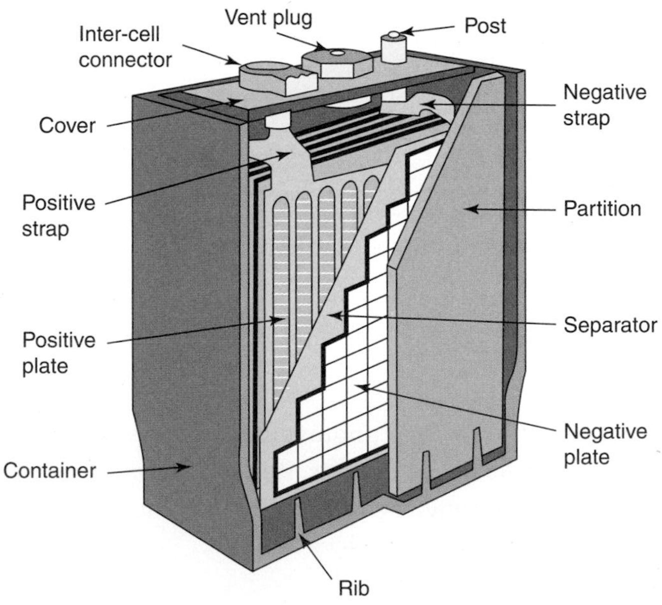

The two main types of deep-cycle batteries that are used in solar PV systems are lead-acid and alkaline batteries. Lead-acid batteries (Figure 4-15) can be divided into two groups:

- Liquid vented
- Sealed (VRLA—valve-regulated lead acid)

Liquid-vented lead-acid batteries consist of positive and negative plates that are immersed in an **electrolyte** solution of sulfuric acid and water. They closely resemble an automotive-type battery. The water in the electrolyte solution is lost when the battery vents its waste gases, and therefore needs to

Inter-cell connector
Vent plug
Post
Cover
Negative strap
Positive strap
Partition
Positive plate
Separator
Negative plate
Container
Rib

© CENGAGE LEARNING 2012

Figure 4-15

A cut-away of a standard lead-acid battery.

be replaced periodically. Just like with the automotive battery, the liquid-vented type is susceptible to reduced power output in cold conditions. Conversely, higher ambient temperatures will shorten the battery's productive lifespan.

Sealed lead-acid batteries have no vents, and therefore have no access to the electrolyte solution. However, they do contain a valve that allows excess pressure to escape in the event of an overcharging situation, hence the name "valve regulated lead-acid" (VRLA). It is important to remember that sealed batteries must be charged at a lower amperage rate to prevent excess gases from damaging their cells.

Alkaline batteries can also be classified into two groups:

- Nickel-cadmium
- Nickel-iron

Both the **nickel-cadmium** (Figure 4-16) and **nickel-iron** batteries also contain positive and negative plates that are immersed in an electrolyte solution. However, this solution is made up of potassium hydroxide. The main difference between lead-acid and alkaline batteries is not so much performance as it is cost. Alkaline batteries are more expensive than lead-acid, but will have a longer lifespan, and they can be discharged completely without adverse affects. However, lead-acid batteries cannot be discharged more than 50% without concerns of shortening their usefulness. For these reasons, alkaline batteries are typically found only in commercial and industrial applications.

Battery capacity is rated in **amp-hours** (Ah). This capacity is based on the amount of power required to operate the designated loads, and on how many days of stored power will be needed for emergency situations. Most manufacturers rate the capacity of their batteries in amp-hours. For instance, a 50-Ah battery will deliver 1 amp of current for 50 hours before it is considered fully discharged. If more current is needed for storage capacity, additional batteries can be connected

Tech Tip
Using Charge Controllers

Lead-acid batteries must incorporate the use of a **charge controller** to prevent overcharging and excessive discharging. These controllers work by monitoring the battery's voltage. This voltage will rise and fall as the battery is charged and discharged. Overcharging will shorten the battery's life because of a loss in its electrolyte.

Figure 4-16

A cut-away of a standard nickel-cadmium (NiCad) battery.

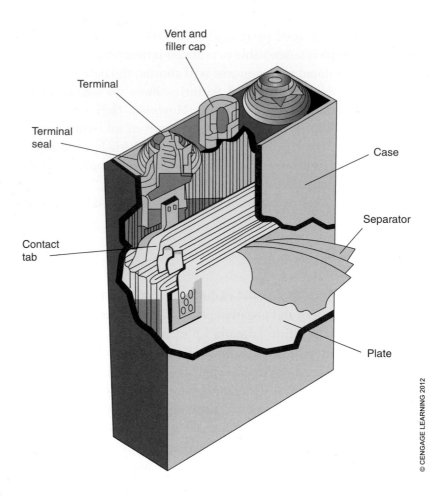

© CENGAGE LEARNING 2012

in a parallel configuration. For instance, two 12-volt, 50-Ah batteries can be wired in parallel to produce 100 amp-hours at 12 volts.

PV Controllers

Photovoltaics require some type of control device to prevent the batteries from becoming overcharged by the solar panels. In this respect, PV controllers act as voltage regulators. Conversely, some controllers work to prevent the batteries from becoming overly discharged by the circuit load as well. In any event, one of the main purposes of the PV controller is to prolong the life of the batteries. PV charge controllers continuously monitor the battery voltage and reduce or stop the charging current when the voltage is at its proper level (Figure 4-17).

They are also rated by the amount of amperage that they can withstand. The National Electric Code requires that controllers be capable of withstanding up to 25% of over-amperage for a limited amount of time. This allows the controller to survive the occasional over-amperage effect caused by excessive sunlight. A PV controller also prevents reverse current flow during the nighttime hours. This situation occurs when a small amount of current flows backward through the PV modules, causing the batteries to discharge.

Figure 4-17

Solar controllers continuously monitor the battery's voltage.

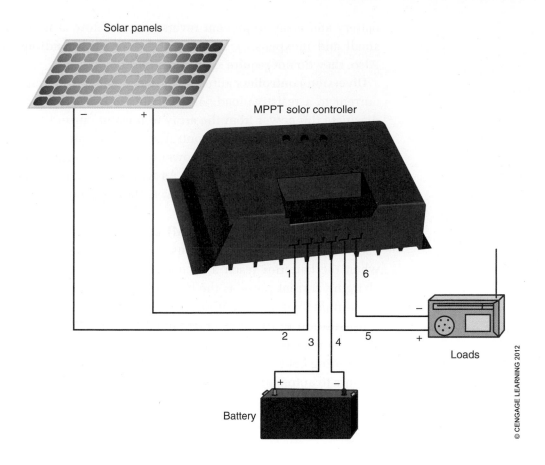

Solar panels

MPPT solor controller

1　6

2　3　4　5

Loads

Battery

© CENGAGE LEARNING 2012

There are four different types of PV controllers currently available:

- Shunt controllers
- Single-stage series controllers
- Diversion controllers
- Pulse-width-modulation (PWM) controllers

Shunt controllers are designed for smaller systems. They prevent battery overcharge by short-circuiting, or "shunting," the PV modules when the batteries are fully charged. The shunt controller circuitry monitors the battery's voltage, and then switches the current flow coming from the solar panels through a power transistor when the fully charged setpoint has been reached. This transistor acts like a resistor by converting the excess amperage coming from the solar array into heat. Shunt controllers are simple in design and inexpensive. They are often completely sealed, but must be exposed to open air to provide ventilation for heat removal.

Single-stage series controllers are used to switch the solar array off when the battery voltage reaches a predetermined value called the charge termination setpoint (CTSP). This prevents the batteries from becoming overcharged. When the battery charge drops, another preset value called the charge resumption setpoint (CRSP) causes the solar panels to become reconnected with the batteries. As with shunt controllers, single-stage controllers incorporate relays to break the circuit between battery and array to prevent reverse current flow at night. These controllers are small and inexpensive, but have a greater load-handling capacity than shunts. Also, they do not require significant ventilation.

Diversion controllers automatically regulate current charge by diverting excess current to a resistive load. As the batteries reach full charge, the load resistors dissipate the power from the array to prevent current flow to the batteries. What is unique about this type of controller is that it can be used with other **alternative energy** sources such as wind power or hydro power. They can even be used with a combination of these sources. As with shunt controllers, diversion controllers require proper ventilation; however, they are not designed to prevent reverse current leakage at night.

Pulse-width-modulation (PWM) controllers are the most common controller used with residential applications. These controllers shut off the charging current when the batteries reach a predetermined setpoint by gradually decreasing the charging current pulse as the battery voltage rises. Most PWM controllers have a built-in method of preventing reversing current leakage at night.

In addition to the above-mentioned features, PV controllers may also include such features as overdischarge protection, load management, indicator lights for monitoring controls, system performance meters, and temperature compensation sensors to automatically change charging voltage (Figure 4-18).

PV Inverters

Solar photovoltaic modules will only generate DC power. Conversely, alternating current is the standard power that is used for almost all residential and commercial operations. For this reason, it is essential that the DC power that is generated by the solar PV system be converted to AC power. This is where an **inverter** fits in. The fundamental purpose of a solar PV inverter is to convert the DC electricity from a solar PV system to AC electricity so that it may be used to power alternating-current loads. In addition, inverters can be used so that the power produced by the solar array can be fed back onto the main electrical grid. An inverter works by running the DC input power into two or more power-switching transistors. These transistors are rapidly turned on and off, abruptly switching the polarity of the DC electricity from

Figure 4-18

Solar controllers include such features as indicator lights for monitoring controls, system performance meters, and temperature compensation sensors.

COURTESY OF CHINA SOLAR FLASHLIGHT

Figure 4-19

Solar inverters such as this convert direct current from the array to alternating current.

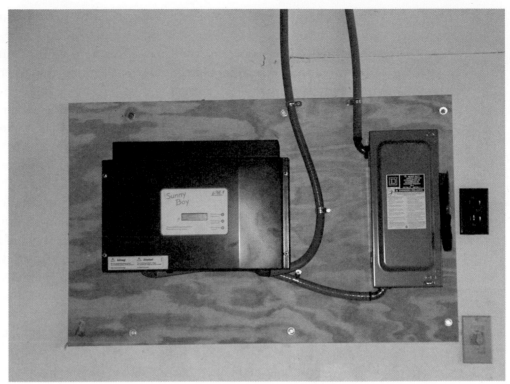

PHOTO COURTESY OF MARY MCGRAW-BIGELOW

positive to negative at approximately 60 times per second. This method creates a square waveform, and a somewhat crude form of alternating current. The power is then passed through a transformer to increase the voltage (Figure 4-19).

Inverters may be classified as grid-tied, grid-tied with battery backup, and stand-alone. The first two types are used with PV systems that are connected directly to the main electrical grid. Stand-alone inverters are also referred to as off-grid types and are used for independent, utility-free power systems. These types are more appropriate for remote photovoltaic installations.

The other category that inverters are classified in deals with the type of waveform they produce. The three most common waveform types are square wave, modified square wave, and sine wave (Figure 4-20). A sine wave is the graph or curve used to describe the characteristics of alternating current and voltage.

Square wave inverters are the simplest type of inverters. As mentioned, these types of inverters simply run the DC power through switching transistors and then into a transformer. They provide little output control and offer limited surge capability. Square wave inverters are limited for use with small resistive heating

Figure 4-20

Three common waveforms that are produced by solar inverters.

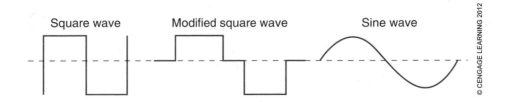

Square wave   Modified square wave   Sine wave

© CENGAGE LEARNING 2012

loads and incandescent lighting. They are not typically used for residential applications.

Modified square wave inverters are a little more sophisticated because they use silicon-controlled **rectifiers** to convert the DC input into alternating current. In a modified square wave, the current and voltage move in steps from peak to valley along the graph. They are more capable of handling higher-voltage surges and produce an output with much less harmonic distortion. Harmonic distortion is the result of a disruption in the voltage and current waveforms within the electrical circuit, and can have an adverse effect on electrical loads, especially those with sophisticated circuitry.

Sine wave inverters are used whenever sensitive electronic loads are involved that require a high-quality waveform. They are used quite frequently with residential applications. The advantages of using this type of inverter over others is that sine wave inverters are specifically designed to produce an output with very little harmonic distortion, and they are resistant to high-voltage surges. This makes them essential for use with sensitive electronic equipment. Grid-tied solar applications require that a sine wave inverter must be used.

Some features that should be applied to all inverters include the following:

- They should maintain an efficiency rate of approximately 90% or better.
- The inverter should maintain this efficiency when there are no loads operating.
- The inverter should maintain a frequency output of 60 Hz over a wide variety of input conditions.
- The inverter should be easy to maintain in the field.
- The inverter should be low maintenance and highly dependable.
- The inverter should be lightweight and easy to install.

APPLICATIONS FOR PHOTOVOLTAIC SOLAR PANELS

Once there is a clear understanding of the fundamentals of solar photovoltaic systems, these fundamentals can be applied to the sizing, installation, and maintenance of PV systems. This chapter will address proper practices plus discuss site and load analysis, as well as look at wiring fundamentals for solar PV systems.

PHOTOVOLTAIC SYSTEM CONFIGURATIONS

As discussed in Chapter 4, there are two main classifications for solar photovoltaic system: grid-connected and stand-alone systems. There are different configurations of PV systems that fall within these two classifications. In addition, PV systems can be interconnected with other forms of alternative energy to create a hybrid system (Figures 5-1 and 5-2).

In its simplest configuration, a solar PV module can supply enough direct-current (DC) power to satisfy the needs of a given application, such as a well pump, highway sign, and even an air conditioning unit (Figure 5-3).

With this type of configuration, the PV module is connected to a charge controller before the electricity enters the battery pack. From there it is connected directly to the DC load (Figure 5-4).

COURTESY OF DONALD STEEBY

Figure 5-1

Solar energy can be interconnected with wind energy to create a hybrid system.

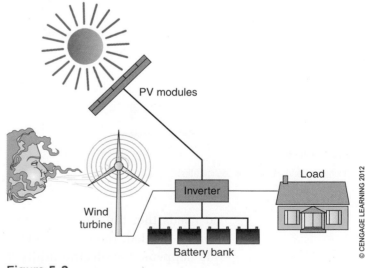

© CENGAGE LEARNING 2012

Figure 5-2

A potential layout for a hybrid system.

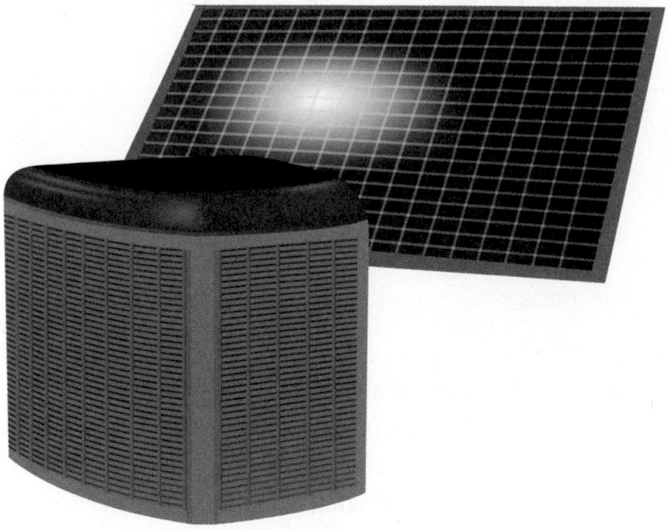

Figure 5-3

Solar panels can even be used to power air conditioning units.

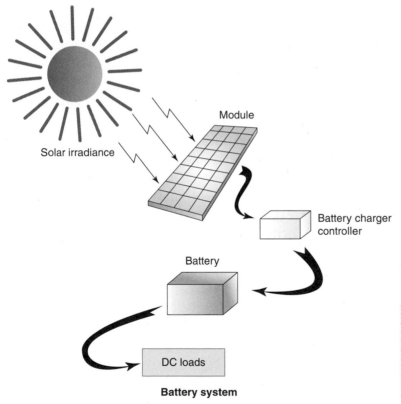

Figure 5-4

A simple stand-alone solar configuration.

Grid-Connected Photovoltaic Systems

Grid-connected systems are designed to function in parallel with the existing electrical grid. As shown in Figure 5-5, the PV module is directly connected to the inverter, which converts the DC power coming off the array into useable alternating-current (AC) power. This conditioned electrical power is consistent with the voltage and frequency requirements of the existing grid. From there, either the inverter or another isolation device automatically shuts down the connection between the PV system and grid, if the main electrical grid is not in service. This device is essential to the safety of utility line workers, who could be fatally injured if any solar-generated power is placed onto the electrical grid while it is under repair. The PV-generated electrical power is then fed either to the electrical loads of the building or is backfed onto the main electrical grid when the solar array output is greater than the on-site electrical demand.

Most building owners who choose this type of solar PV configuration take advantage of net metering. **Net metering** is the process where the building owner receives full value for the power that is produced by the solar photovoltaic system. By the use of net metering, a home or business can offset the cost of its electric bill with any excess electricity that is produced. Here is how it works: when the solar PV system produces electricity, the power is first used to meet any electrical needs that the structure requires. When more electricity is produced than what is needed, the excess power is fed back onto the main **utility grid**. When this happens, the electric utility meter will run backward and the customer receives credit for the additional energy that is placed on the grid. At the end of the

Figure 5-5

A grid-connected solar configuration.

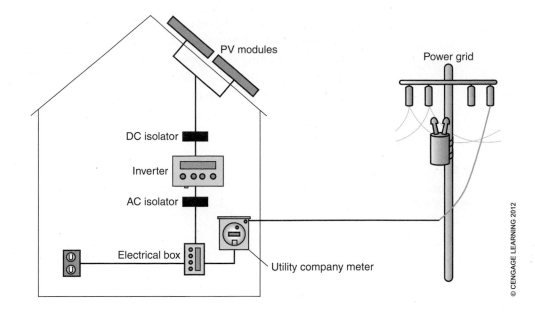

© CENGAGE LEARNING 2012

billing period, the utility company credits the customer with the net amount of kilowatt-hours produced at the wholesale power rate, unless the customer uses more electricity than the solar PV system generates; in which case the customer pays the difference. Under federal law (PURPA, Section 210), local utilities must allow independent homeowners and businesses to interconnect with the utility grid, and the utility company must purchase any excess electricity that is generated. If the homeowner or business owner is in an area where net metering is not allowed, the utility will install a separate meter and purchase the excess electricity at a wholesale price, usually much lower than the retail price. In some states, excess power credits are carried over to the next billing period for up to 1 year. One of the benefits of net metering is that it allows homeowners and businesses to receive the full value of their PV system without having to install a battery storage system. In the case of net metering, the main power grid is the customer's backup system. On the other hand, if the main grid goes down when there is no sunshine, the customer risks being without power for an extended period of time. It is important that the homeowner or building owner check with the local electrical utility company to find out the policy regarding excess electricity that is placed back onto the grid. Net metering is currently being offered in more than 35 states.

Stand-Alone Photovoltaic Systems

With a stand-alone PV system, the building owner operates independently of the electrical utility grid. This type of system is designed to be sized to supply enough electricity to satisfy all of the AC electrical loads of the given structure. The stand-alone system may exclusively use a solar PV system as its source of power or may be interconnected with other sources of electricity, such as a wind turbine or fuel-driven generator (Figure 5-6). Figure 5-7 shows what a stand-alone residence might look like where solar and wind power sources are integrated together.

In most stand-alone systems, a charge controller links the solar array to the battery pack. The battery pack must be sized properly in order to maintain enough electrical current to the structure during prolonged periods of no sun. From there, the electrical power is fed into an inverter before it enters into the building through the AC service panels. The stand-alone system can also take advantage of net metering as a means of taking full advantage of the solar PV array (Figure 5-8).

Figure 5-6

A solar configuration can integrate with an electrical generator.

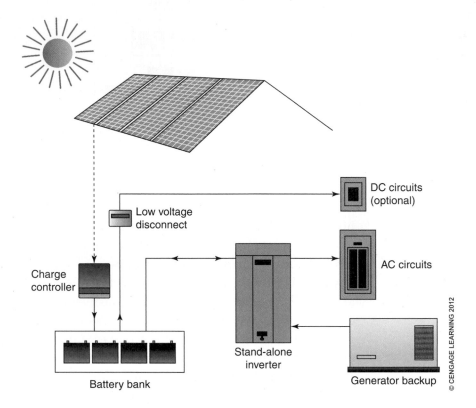

© CENGAGE LEARNING 2012

Figure 5-7

How a residence might look when solar and wind energy are integrated together.

© CENGAGE LEARNING 2012

Figure 5-8

A typical configuration for a solar PV stand-alone system.

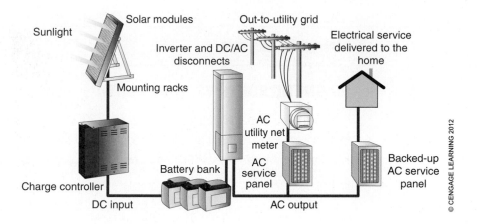

UTILIZING SOLAR RESOURCES

It is important to understand the relationship of the sun's angle to the earth in order to effectively install a solar photovoltaic array. Much of the information needed to understand this relationship has been covered in Chapter 3 under the heading "Solar Angles." This information applies to the orientation of solar arrays used for photovoltaic applications as well as for solar thermal storage.

As stated earlier, the total amount of energy from the sun that strikes the earth's surface on a clear day is about 1,000 watts per square meter. The term for this energy is known as solar **insolation**. The amount of solar insolation at a given point will vary depending upon geographic location, atmospheric conditions, and the amount of obstructions at the location of the solar array. The amount of solar insolation that is calculated at a given area can also be used as a tool for sizing the array. For example, the maximum amount of sunshine at a given location can determine the number of peak sun hours. **Peak sun hours** are equal to the number of hours per day when the solar insolation is equal to 1,000 watts per square meter. An area that receives 6 peak sun hours is equal to receiving 6 kWh per square meter. Peak sun hours are usually used to determine the amount of energy received during total daylight hours. This data is made available for any given location by the National Renewable Energy Laboratory (NREL) and can be found at http://nrel.gov.

The other factors that affect the amount of solar insolation for a given area are determined by the sun's orientation in the sky and its tilt angle. In the northern hemisphere, the orientation of the sun in relationship to **true south** is called the azimuth. The azimuth measures the angle in degrees east or west of true south. True south is not the same as the magnetic compass direction south, or **magnetic south**. True south is simply the midway point of the sun between where it rises in the east and sets in the west for any given time of year. The best orientation for a solar array is to face true south or 0 degrees azimuth (Figure 5-9).

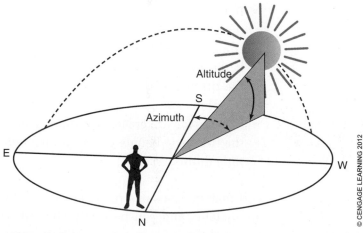

Figure 5-9

The azimuth of the sun measures the angle in degrees east or west of true south. The sun's tilt angle or height above the horizon is considered the sun's altitude.

Things to Know

MAGNETIC SOUTH VS. TRUE SOUTH

If a particular site has a magnetic declination of 20° east, it means that true south for this location is 20° east of magnetic south. If a compass needle is at 360° when pointing north, true south for this location is indicated on the compass at 160°, rather than 180°.

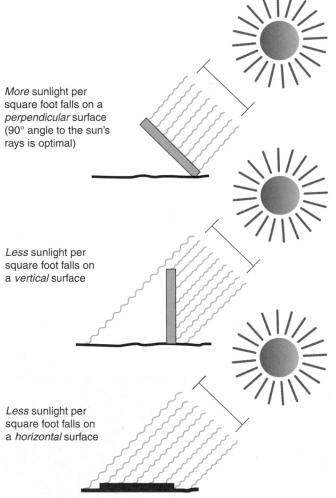

More sunlight per square foot falls on a *perpendicular* surface (90° angle to the sun's rays is optimal)

Less sunlight per square foot falls on a *vertical* surface

Less sunlight per square foot falls on a *horizontal* surface

Figure 5-10

Seasonal changes to the sun's orientation must be considered in order to optimize the solar PV system's performance.

© CENGAGE LEARNING 2012

The sun's tilt angle or height above the horizon is considered the sun's altitude. This factor is measured in degrees above the horizon. When the sun is at the highest point for a given day, it is said to be at its solar noon. This angle or altitude will differ throughout the year due to the earth's orbit around the sun with its tilted axis. Solar PV arrays work best when the sun's rays are shining perpendicular to the solar cells. Therefore, seasonal changes to the sun's orientation must be considered in order to optimize the solar PV system's performance. If the building's loads change on a seasonal basis, then the optimum tilt angle of the array must be taken into consideration (Figure 5-10).

Following are some rules to follow when considering these factors:

- If the building's loads are consistent throughout the year, the tilt angle equals the location's latitude.
- If the loads are greater in wintertime, the tilt angle equals the location's latitude plus 15°.
- If the loads are greater in summertime, the tilt angle equals the location's latitude minus 15°.

Fortunately, the sun's path across the sky is predictable and can be easily charted in order to calculate the maximum performance of the PV system. In addition, some newer solar PV arrays can be controlled on a pole-mounted, articulating axis to take advantage of the sun's optimum orientation at any given time of year. The NREL website also offers an excellent resource for calculating the annual energy production of a grid-tied solar photovoltaic system called "PVWATTS." This program allows the solar designer to input the tilt and orientation angles for a given system, and then calculates the corresponding electrical production based on these values.

Site Analysis

It is important to gather correct and accurate data from the given site where the solar PV system is to be implemented, in order to achieve a successful installation. The data collected will be even more critical if the system is to be considered stand-alone. In the process of data collection and interpretation, the following steps will be involved:

- Gathering solar insolation data
- Determining the solar design month
- Determining solar access
- Identifying shading obstacles

When sizing a solar PV system that will meet the electrical load demands of a given structure, it is recommended that the amount of solar insolation for the given area first be determined. Many locations around the world have published data regarding the amount of available solar insolation that will be significant for designing a solar PV system. These published data are made available through meteorological sites such as the National Oceanic and Atmospheric Administration (NOAA), the NREL, or through manufacturers of solar photovoltaic systems. This information should include the solar insolation available at different tilt angles and through different tracking options.

Once the solar insolation levels for the given area have been determined, the next step is to calculate the proper design month. The solar design month is that month of the year with the lowest level of solar insolation, and if possible the highest electrical load demand. When the proper month has been determined, the solar array should be oriented so that the tilt angle yields the maximum amount of insolation for that month. By doing this, the system will be designed to meet the building's maximum electrical load and keep the battery pack fully charged under the worst-case scenario.

Another method is to design for the month with the highest electrical demand. The procedure for this method is to first calculate the building's electrical demand for each month, then divide this number by the respective month's average solar insolation. The month that corresponds to the largest design electrical demand should be considered the system's solar design month.

Determining solar access involves finding a site with adequate year-round exposure to sunlight. The period during the day with the maximum amount of sunlight available is considered to be "solar time," and this may differ from the actual time of that geographic location in relationship to the country's time

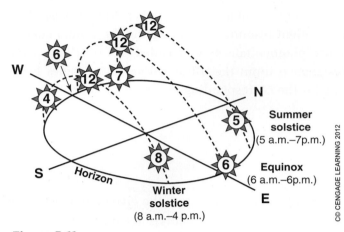

Figure 5-11

The solar time will vary according to the season of the year.

zones (Figure 5-11). For instance, solar noon is when the sun is at its highest peak in the sky for that time of year, and that time may not necessarily be at 12:00 p.m. Sun charts are available for specific latitudes from energy agencies and solar suppliers that can help assist the designer in determining the solar path for a particular area. As mentioned in Chapter 3 under the heading "Solar Angles," the University of Oregon Solar Monitoring Laboratory website can also generate a solar path diagram for any given location and: http://solardat.uoregon.edu/SunChartProgram.html.

Finally, it is imperative that the solar installation site be free from shading obstacles. Even small amounts of shading from such things as trees, buildings, and hills can greatly reduce the performance of the solar array. Minimizing shading is much more important in the design of a solar photovoltaic system than it is in designing a solar thermal storage system. As a rule of thumb, the solar array should be free of any shading from 9:00 a.m. through 3:00 p.m. This optimum solar collection time frame is known as the "solar window" (Figure 5-12).

Shading is even more of an issue during the winter months when the sun's altitude is lower in the sky, making shadows longer. As mentioned in Chapter 3, one device that is used to determine the effects of shading is called a solar pathfinder. This device is placed at the location where the amount of shading is to be evaluated. After the pathfinder has been placed in the proper orientation, its clear hemispherical dome projects the reflections of nearby objects onto a special chart that will indicate when the desired location is in the shade.

Figure 5-12

The optimum solar collection time frame is known as the solar window and takes into account any shading factors.

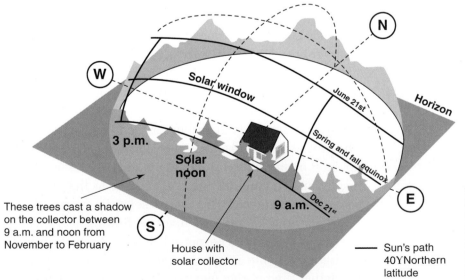

Load Analysis

An important part of the overall plan when designing the proper solar PV system should be to include an audit and analysis of the building's electrical load requirements. This information can form the basis from which to properly size the solar array and its peripheral devices. Thoroughly analyzing the energy requirements of the building can lead to a reduction in electrical usage by identifying energy conservation opportunities and reduce the size and cost of the solar PV system installation.

The simplest means of calculating the overall electrical load for a given building, typically on a monthly basis, is to formulate a spreadsheet. This spreadsheet would first list the type of appliance and its listed power consumption, usually in watts. Most manufacturer nameplates list the wattage required for the equipment's load. If this information is not available, the power usage in watts can be calculated by multiplying the supply voltage times the appliance current draw. The current can be measured using a standard clamp-on **ammeter** (Figure 5-13).

Next, estimate the run time of the appliance in hours for a given day. The run time is sometimes referred to as the duty cycle, which is the percentage of time that the appliance is energized.

Once the electrical usage is determined for a given day, this number can be multiplied by the number of days per month. The monthly total can be used in conjunction with the sun's insolation factor mentioned earlier to calculate the size of the solar array needed. Table 5-1 provides a load estimation worksheet.

© ISTOCKPHOTO/ALEKSANDR UGORENKOV

Figure 5-13

Amperage can be measured by clamping the jaws of an ammeter around the conductor.

Table 5-1

Load Estimation Worksheet

Appliance	Quantity	Voltage	Amperage	Watts	Runtime	Watts per Day	# of Days per Month	Monthly Watts Consumed

© CENGAGE LEARNING 2012

Total Watts per Day =   Total Watts per Month =

Other factors that will influence the electrical load analysis for a given building include the types of appliances that are used. Resistive heating appliances such as electric clothes dryers, water heaters, ovens, and space heating devices are not recommended for use with a solar PV system. Powering these devices can be too cost prohibitive for a solar photovoltaic system, especially if the system is to be considered stand-alone. Other types of appliances, such as gas-fired models, should be considered as an alternative.

Tech Tip
Choosing the Right Appliances

The refrigerator is one of the single largest energy loads in the residential home. Because most people choose not to live without a refrigerator, deciding on an efficient model will not only reduce the monthly utility bill, it will also significantly lower the initial cost of the solar PV system by lowering the energy consumption.

Another means of achieving energy conservation is to analyze the building's lighting load. Incandescent lamps can be replaced with compact fluorescent lamps as a way of lowering the total electrical load of the structure (Figure 5-14).

Figure 5-14

Replacing conventional incandescent bulbs with compact fluorescent light bulbs can reduce the lighting load dramatically.

© ISTOCKPHOTO/JON SCHULTE

For commercial applications, the entire incandescent fixture can be replaced with a low-energy fluorescent fixture. In addition, the way the light fixture is controlled can reduce the time that the light is burning, and also lower the overall electrical load. Conventional manual light switches can be replaced with timers, photocells, or occupancy sensors.

PHOTOVOLTAIC SYSTEM WIRING

The proper wiring of the photovoltaic system is one of the most important elements in the installation of the system, and must be thoroughly understood. Furthermore, all wiring of any photovoltaic system must comply with the National Electric Code (NEC), specifically Article 690, which covers solar photovoltaic systems. NEC 690.4(E) states that all electrical wiring, whether it be the installation or service of a PV system, shall be performed by qualified persons. The information presented here references the 2011 National Electric Code book.

Wire Types

The two most common types of electrical wire used are aluminum and copper. Aluminum wire is typically less expensive than copper; however, it does not possess the same conductivity or current-carrying capability as copper wire. Furthermore, aluminum wire is less durable and is not permitted for use with interior residential applications.

Wire is available and sold as either solid or stranded. Stranded wire is made up of many small wires that are bundled together, making it more flexible than solid. Also, solid wire has a greater tendency to break if it is nicked when stripping off the exterior insulation (Figure 5-15).

The insulation that covers the wire is available in many different forms depending upon the application that it is used for. Table 5-2 outlines various types of wire

Figure 5-15

Comparing solid wire to stranded.

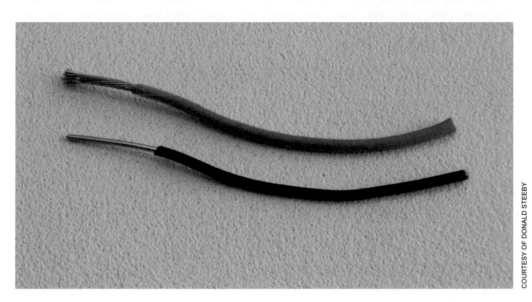

COURTESY OF DONALD STEEBY

Table **5-2**

Various Wire Types and Their Applications

Type	Covering Temp	Max. Provisions	Location Covering	Insulation	Outer
THHN	Heat Resistant Thermoplastic	90°C 194°F	Dry or Damp	Flame Retardant & Heat Resistant Thermoplastic	Nylon Jacket
THW	Moisture & Heat Resistant Thermoplastic	75–90°C 167–194°F	Dry or Wet	Flame Retardant & Moisture & Heat Resistant Thermoplastic	None
THWN	Moisture & Heat Resistant Thermoplastic	75°C 167°F	Dry or Wet	Flame Retardant & Moisture & Heat Resistant Thermoplastic	Nylon Jacket
TW	Moisture Resistant Thermoplastic	60°C 140°F	Dry or Wet	Flame Retardant & Moisture Resistant Thermoplastic	None
UF	Underground Feeder & Branch Circuit Cable-Single Conductor	60–75°C 140–167°F	Service Entrance	Moisture and Heat Resistant	Integral with insulation
USE	Underground Service Entrance Cable-Single Conductor	75°C 167°F	Service Entrance	Moisture and Heat Resistant Non-metallic Covering	Moisture Resistant

© CENGAGE LEARNING 2012

A more complete table can be found in the *NEC® 2005,* Table 310.13.

and the applications in which they are used. Careful consideration must be paid to the type of wire and its specific use to ensure that it is applied properly.

Color coding is another important aspect of wiring the solar PV system. Color coding designations ensure safe and efficient installations, repairs, and troubleshooting. The three most common color coding designations for both AC and DC electrical wiring are:

- **For an ungrounded conductor:** Any color other than green, white, or gray
- **Grounded conductor (neutral or negative):** White or gray
- **Equipment ground:** Green or bare wire

Tech Tip
Color Coding Larger Conductor Wire

Larger conductors over #4 **American Wire Gauge (AWG)** typically use black as a color code. To designate different conductors that are in a bundle, different colored electrical tape is used on the ends of each wire.

Cables and Conduit

Electrical cabling is described as two or more insulated conductors that are bundled together into one insulated jacket. A typical application for this type of wire would be for low-voltage control wiring, such as for a thermostat. As with individual conductors, this jacketing is rated for a specific use. For instance, the cable may be resistant to moisture, UV light, heat, or chemicals. A more extensive definition of proper cabling applications can be found in the National Electric Code.

Conduit may be metal or plastic pipe that contains the individual wires to create a protective enclosure. Conduit is especially used in areas where the wiring may be subjected to damage or to the elements. For indoor applications, electrical metallic tubing (EMT) may be used. EMT is lightweight, durable, and price competitive. For outdoor use, polyvinyl chloride or PVC conduit is typically used, especially for underground applications. Liquidtight conduit is also used for outdoor applications, especially to connect modules to a junction box. Consult the NEC to determine the maximum number of conductors and size of wire that can be run within a given conduit size.

Wire Sizing

There are two important considerations when selecting the proper size of wire for a project:

- Amperage capacity
- Voltage drop

Amperage capacity, or **ampacity**, refers to the current-carrying ability of the conductor. The larger the wire is, the greater its capacity for carrying current. Overheating will result if the wire is undersized. This situation can cause the insulation to melt or even cause a fire. Table 5-3 gives the maximum amperage capacity for a given wire based on its size.

Note that the amperage capacity is lower if the wire is run through conduit or cable.

Wire is sized according to American Wire Gauge (AWG). With this system, a larger wire size has a designated smaller number. Conversely, the smaller the wire size, the larger the number. The smallest size wire typically used for construction purposes is #24 AWG. Regarding larger-sized wire, after #1 AWG, the designated symbol /0 is used, all the way up to #4/0 wire.

The first step in sizing wire for a particular circuit is to first determine the maximum current load. Once the maximum current has been determined, this number is multiplied by 125%. This is done so that the conductor is limited to no more than 80% of its rated carrying capacity. When sizing the wire between the PV array and the battery pack or controller, use the "short-circuit" current of the PV module multiplied by the number of modules that are wired in a parallel configuration. This figure is then multiplied by an additional 125% (or 156% total) due to the fact that the NEC requires an additional safety factor for wiring that connects the PV

Table 5-3

Amperage Capacity of Copper Wire

AWG	In Conduit or Cable		Single Conductors in Free Air	
	UF, THW	USE, THWN	UF, THW	USE, THWN
14	15	15	20	20
12	20	20	25	25
10	30	30	40	40
8	40	50	60	70
6	55	65	80	95
4	70	85	105	125
2	95	115	140	170
1/0	125	150	195	230
2/0	145	175	225	265
3/0	165	200	260	310
4/0	195	230	300	360

© CENGAGE LEARNING 2012

For a more complete table see *NEC®2005*, Table 310.16 and Table 310.17.

array to the battery pack or to an inverter in a system not using batteries. For more information on circuit sizing and current, reference NEC 690.8.

Voltage drop is the second consideration when wire sizing. Voltage drop is a function of the resistance of the wire, the size of the wire, and its overall length. Obviously, a longer wire run will result in a greater voltage drop across the circuit. This issue can be critical, especially when electric motors are involved. The solution to this situation is to correctly size the wire based on the length of the circuit and to try to keep the wire runs as short as possible. A standard design practice is to maintain the voltage loss of a given circuit between 2% and 5%. Once again, the NEC provides tables that will assist in calculating the proper voltage drop allowed based on wire size and length of circuit.

Over-Current Protection

Each circuit in the solar photovoltaic system must be protected from amperage that exceeds the wire's amperage capacity. The two main types of circuit protection are:

- Fuses
- Circuit breakers

Each of these devices will interrupt the circuit, causing it to open and stopping the current flow.

Fuses contain a strip of metal that has a higher resistance than the conductor in the circuit (Figures 5-16 and 5-17). This strip will melt when the current exceeds its rated amperage. Fuses are a one-time device that must be replaced after they have blown. Common causes of fuse failure are overloaded circuits and short circuits caused by faulty wiring, faulty grounding, or equipment failure. NEC section 690.16 contains guidelines for properly fusing circuits.

Circuit breakers can function as a switch as well as a circuit protector. Most modern wiring installations use circuit breakers instead of fuses for circuit protection (Figure 5-18). Circuit breakers use two different types of circuit protection. One type uses a **bimetal strip** that, when heated by a current overload, will cause the breaker to trip. The other type uses a magnetic coil that will open the circuit when there is a quick inrush of amperage.

It is important to note that breakers used for AC circuits are not suitable for DC circuits, unless they are specifically rated for that purpose. Other rules regarding the use of over-current protection include:

- Every underground conductor must be fuse/breaker protected.

Figure 5-16

A plug-type fuse.

© CENGAGE LEARNING 2012

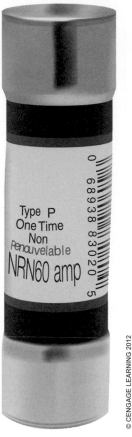

© CENGAGE LEARNING 2012

Figure 5-17

A cartridge-type fuse.

© ISTOCKPHOTO/DON NICHOLS

Figure 5-18

A circuit breaker.

- Each power source must have an over-current device.
- If the rated ampacity of a wire falls between two fuse/breaker sizes, the next larger over-current device shall be used.

Figure 5-19 shows a solar PV system with the proper placement of over-current protection.

Things to Know

THE TASK OF THE OVER-CURRENT DEVICE

It would seem that the main task of the fuse or circuit breaker would be to protect the equipment from damage. However, the primary purpose of an over-current device is to protect the wire from overheating and potentially causing a fire.

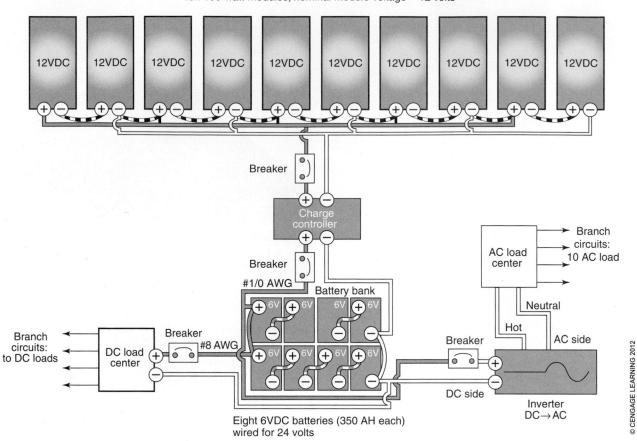

Ten 100-watt modules; nominal module voltage = 12 volts

Figure 5-19

The proper placement for over-current protection on a solar PV system.

GROUNDING REQUIREMENTS

Proper **grounding** of the solar photovoltaic system is one of the most essential yet complex steps in completing the wiring installation. Grounding is defined by the National Electric Code as connecting the equipment or system to the earth or to a metallic conducting body that serves as earth.

Grounding accomplishes several important steps:

1. It limits the amount of voltage that the equipment is subjected to during a lightning strike or electrical surge.
2. It stabilizes the voltage within the circuit and provides a common reference point with the earth.
3. It provides a path for the electricity to take in the event of an over-current condition within the electrical circuit.

Electricity follows the path of least resistance. Without proper grounding, a human could potentially become that pathway for the electricity to take.

The result can be severe shock that could result in injury or even death. For this reason, the National Electric Code requires that for all two-wire solar PV systems over 50 volts, one of the DC current-carrying conductors shall be grounded. This also requires that the negative conductor in the DC system be bonded to ground at one single point in the system. The negative conductor is the white or gray wire. It is suggested that on a solar PV system, the grounding connection point be as close as practical to the array or panel to better protect the system from a voltage surge due to lightning. This procedure is known as **system grounding**.

In addition, **equipment grounding** is necessary in all solar PV systems as well. Equipment grounding requires that every non-current-carrying metal portion of the installation be connected to a continuous wire that is connected to ground. This includes bonding every metal box, receptacle, equipment chassis, appliance frame, and PV panel mounting. This ground wire must never be fused, switched, or interrupted in any way. Essentially, by following the proper grounding procedures, both the AC and DC circuits will be properly grounded.

Most solar PV systems will use a minimum #6 AWG bare copper wire that is connected from the system equipment to the grounding rod, known as the grounding electrode. This electrode is a rod made of iron or steel that is at least 5/8″ thick, and is to be driven into the ground at least 8 feet deep. Figures 5-20 and 5-21 illustrate examples of proper grounding for both stand-alone and grid-tied solar PV systems. Always follow NEC requirements or consult a licensed electrician before attempting to wire a solar photovoltaic system.

Figure 5-20

Proper grounding for a stand-alone PV system.

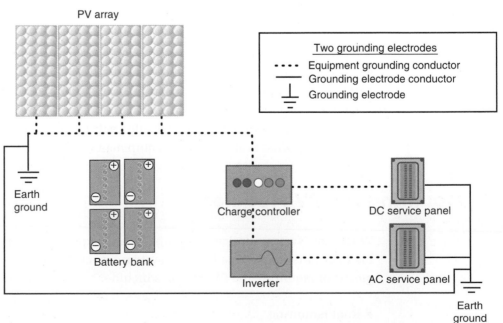

© CENGAGE LEARNING 2012

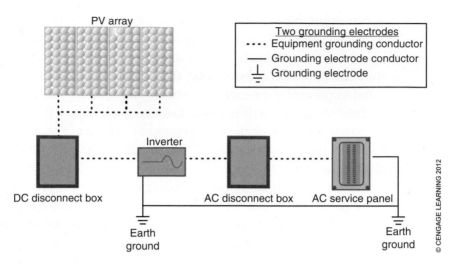

Figure 5-21

Proper grounding for a grid-tied PV system.

INSTALLING PHOTOVOLTAIC SYSTEMS

The installation procedure for a solar photovoltaic system consists of the following steps:

- Site evaluation
- Mounting the PV modules
- Battery, controller, and inverter installation
- System wiring

Site Evaluation

Site evaluation is an important step because each site possesses its own set of unique challenges. The installer should first review the installation site to determine a number of items: whether the area has sufficient sunlight for the PV system to work effectively, whether the array will be roof mounted, the structural soundness of the roof, the electrical load of the building, the location of existing electrical service, and whether the system will be stand-alone or grid-connected.

Tools that should accompany the installer or designer to the job site include: paper and pencil, tape measure, digital camera, inclinometer to measure the roof slope, and flashlight. By performing a thorough site analysis and evaluation, the designer can better decide what type and size of equipment will best suit the customer's needs.

Mounting the PV Modules

When the system design is complete and equipment has been specified, the installer must determine the best system for mounting the modules. The various means of mounting the PV modules include:

- Roof mounting
- Pole mounting

- Ground mounting
- Track mounting

There are several different techniques when roof mounting the solar array. These include mounting the modules directly to the roof (Figure 5-22). As was discussed in Chapter 3 under the heading "Mounting the Solar Panel," several factors need to be considered when mounting solar modules in this manner: make sure that the roof has the structural integrity to support the modules; mount the support brackets directly to the roof trusses or rafters; ensure that the roof is properly sealed and flashed to avoid possible leaks. Another consideration when mounting modules directly to the roof is air circulation. Direct mounting does not always allow for air circulation underneath the PV modules. This can result in higher operating temperatures and a decrease in the power output of the array.

Figure 5-22

A solar array mounted directly to a roof.

To avoid this situation, and to make the modules more accessible for maintenance, the installer may opt for the use of mounting brackets (Figure 5-23). Some rack mounting systems are adjustable, so the system can be adjusted to take advantage of optimum sunlight. If the modules are to be mounted on brackets to maintain the proper angle to the sun, be aware that high winds can be a factor in determining the structural soundness of the mounting hardware.

Pole-mounted arrays are bolted directly to a vertical pole that is securely and permanently placed in the ground (Figure 5-24). When roof mounting is not an option, a pole mount may be the answer. This technique can take advantage of optimal sunlight by allowing the operator to adjust the pitch and angle of the array

Figure 5-23

A roof-mounted solar array using mounting brackets.

Figure 5-24

A pole-mounted solar array.

on a seasonal basis. Local conditions will determine the size and strength of the pole, so that it will hold up under inclement weather. Most solar suppliers can specify the size and strength of the pole to be used.

Ground-mounted systems use a frame and bracket system that is bolted together and mounted on secure footings (Figure 5-25). As in the pole-mount scenario, ground-mounted footings are secured to the ground for added support. This option can also take advantage of seasonal array adjustments for optimal performance. Soil conditions should be evaluated with ground-mounted systems to ensure there is load-bearing capacity for the weight of the array and its mounting hardware. Standard support structures include 4-, 8-, and 12-module brackets for site fabrications.

Track-mounted solar arrays offer an added advantage to pole mountings (Figure 5-26). These devices utilize motors powered by their own integrated solar panels to rotate the array and literally track the sun's path across the sky. There are single-axis trackers that follow the sun's azimuth, and dual-axis trackers that follow both azimuth and altitude for maximum array performance and power output. Solar tracking arrays require a firm foundation due to their additional weight. Typically, a 4- to 6-inch-diameter base set in concrete is required for support. Trackers are typically 25% to 30% more efficient than conventional fixed arrays; however, they are significantly more expensive.

Figure 5-25

A ground-mounted solar array.

© ISTOCKPHOTO/VINICIUS RAMALHO TUPINAMBA

Figure 5-26

A solar array using tracking capability.

PHOTO COURTESY OF MARY MCGRAW-BIGELOW

Tech Tip

Solar Track-Mounted Systems

Systems that require the largest electrical demand in summertime are much more ideal candidates for a solar pole-mounted tracking system, because they can take advantage of the longer daylight hours to increase electrical production and system collection potential.

Battery and Inverter Installation

The procedure for battery installation is not complicated. Simply connect the terminals in either a series or parallel configuration, according to the wiring diagrams. Where the battery installation is critical is in the handling and storage of the batteries themselves. Batteries must be protected at all times, including while they are being transported and while charging them before installation. When pre-charging the batteries, be sure not to overcharge them, and keep them away from open flames, which could cause an explosion from the hydrogen gas that may have escaped from the vent caps.

The battery pack must be installed in a safe location that protects them from the environment yet allows for ease of service. Usually a vented battery box should be provided that is corrosion resistant and insulated. Other items to consider when housing batteries are:

- Protecting them from freezing temperatures
- Providing ventilation to prevent the buildup of explosive gases
- Securing the battery enclosure so it is theft deterrent, yet easily accessible
- Ensuring the enclosure is corrosion resistant
- Ensuring that the area where the batteries are being stored has a strong enough base to support the weight

Inverters and charge controllers must be installed according to manufacturers' specifications. Follow these guidelines to ensure that controllers and inverters are installed properly:

- Protect the equipment from dust and dirt, overheating, and rough handling. These types of equipment usually contain solid-state circuitry that is susceptible to contamination and to static electricity.
- Provide a suitable mounting location that allows for ease of repair and does not interfere with other electrical hazards.
- Always use correct wire sizes and proper terminal fasteners when installing.
- Utilize proper over-current protection and safety disconnects between the batteries and the inverter, or between the charge controller and the inverter.

System Wiring

Once the major system components have been installed, they need to be wired together. Most of the wiring requirements for a correct solar photovoltaic system installation have been covered earlier in this chapter. There are, however, basic wiring practices that should be followed in order to achieve a successful solar installation. These include the proper use of electrical connectors, junction boxes, cabling, disconnects, switches, and receptacles.

Electrical connections are usually either wire to wire or wire to terminal (Figure 5-27). Wire-to-wire connections are made using wire nuts, butt splice connectors, or crimp connectors. When making these types of connections, make sure that the wire has been stripped to its proper length and that connectors are firmly secure. Many times, nuisance service calls are a result of wires not firmly connected together.

All electrical connections must be housed within an accessible electrical junction box. This box must be well secured and have a removable cover. If connections are to be made outdoors, they must be housed in a weather-tight junction box, usually made of PVC material, according to the National Electric Code.

Cable connections require that all individual wires within the cable be securely fastened together. If there is room in an enclosure, the use of a terminal strip makes connecting cables easier and more secure. It also makes the project look neater and more professional.

As mentioned earlier, the NEC requires that each voltage source have its own disconnect. As an added measure, some disconnects will include fuses for over-current protection. Disconnects must be rated for the voltage and current flow through the circuit. Also ensure that disconnects are rated for either alternating current or direct current.

Switches should be included as a means of interrupting the current flow through the circuit, either for use when servicing the system or for overload protection. They should be rated for heavy-duty use and, if applicable, for outdoor use. They need to be located near the equipment so they are readily accessible.

Figure 5-27

Various electrical connectors used in installations.

COURTESY OF DONALD STEEBY

COURTESY OF DONALD STEEBY

Figure 5-28

A ground-fault-circuit-interrupted (GFCI) receptacle.

Receptacles are usually installed near the equipment as "convenience outlets." This is especially true if service is being performed outside, either on the roof or at a remote location. The receptacle must be rated for use with either alternating current or direct current. A convenience outlet will provide the service technician a place to connect power tools and trouble lights. Remember that for outdoor applications, the receptacle must be enclosed in a weather-tight enclosure and be ground-fault-circuit-interrupted (GFCI) rated (Figure 5-28).

PUTTING IT ALL TOGETHER—A REVIEW

Here are the steps that have been covered in planning and installing a solar photovoltaic system:

1. Perform a site analysis.
2. Calculate the building's system load.
3. Determine whether to install a stand-alone or grid-tied system.
4. Size and install the solar array.
5. Size and install the battery pack.
6. Install the charge controller.
7. Size and install the inverter.
8. Wire the system and connect it to the grid.

MAINTENANCE OF THE PV SYSTEM

Surprisingly, solar photovoltaic systems require very little maintenance. Following is a list of items that are typically involved in keeping the system in top working order. The maintenance check should be performed a minimum of twice per year. The ideal time to perform a maintenance check would be around noon on a sunny day.

Figure 5-29

Make sure all hardware connections are secure when maintaining the solar array.

© ISTOCKPHOTO/LISA F. YOUNG

Figure 5-30

Check for any roof leaks when maintaining the solar array.

The Solar Array: Observe the condition of the modules. Check for signs of degradation, such as discoloration, fogged glazing, or water leaks. Tighten all loose nuts and bolts on the mounting racks and clips (Figure 5-29).

Secure any loose wiring beneath the modules. Make sure there is no wiring damage from vermin. Replace any damaged wiring or wire runs. Make sure any roof penetrations are secure and that flashing is securely in place and free of any leaks (Figure 5-30).

Remove any shading restrictions around the array and remove any built-up dirt or debris from the module glass (Figure 5-31).

Batteries: If the system includes a battery pack, this area will most likely require the most amount of time during the preventative maintenance routine. One of the first tasks is to check the battery charge. This can be measured using a standard multi-meter. Be sure to first operate the system for several minutes under load before checking the charge. This will stabilize the voltage and remove any inaccurate surface charges. Disconnect both leads of the battery from the array before taking the voltage readings.

Vented-liquid lead-acid batteries will require the most amount of maintenance because they occasionally require the addition of distilled water to fill the cells. This is due to the fact that lead-acid batteries lose water when hydrogen passes through the vents. These types of batteries may also require their state-of-charge to be checked. This procedure can be performed with

Figure 5-31

Remove any shading obstacles when performing maintenance.

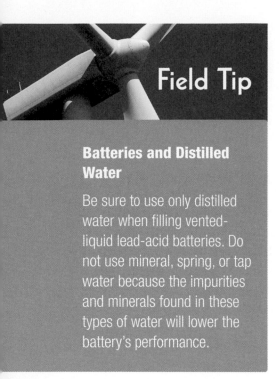

Field Tip

Batteries and Distilled Water

Be sure to use only distilled water when filling vented-liquid lead-acid batteries. Do not use mineral, spring, or tap water because the impurities and minerals found in these types of water will lower the battery's performance.

the aid of a hydrometer. This device is similar to one used to check the antifreeze level on an automobile (Figure 5-32).

The glass float in the hydrometer is calibrated to read the specific gravity of the battery's electrolyte. These devices are calibrated to give a true reading when the ambient air temperature is about 80°F. Use the correction factor chart that accompanies most hydrometers when testing batteries while the temperature is above or below this setpoint. Remember that some new batteries require several cycles before they will give an accurate reading with the hydrometer.

Inverters: Use a voltmeter and DC ammeter to check and record the operating input voltage on the DC side. Also check the AC current level and output voltage. If the inverter has light-emitting diode (LED) indicator lights, be sure that the lights indicate proper operation. If the inverter displays the total kWh, check this number to ensure that it is accurate since the last maintenance update.

Electrical Components: Open all junction boxes and check electrical connections for tightness (Figure 5-33).

Open all disconnect switches and use an ohmmeter to check for proper grounding. An ohmmeter reading of greater than 25 ohms indicates the presence of corrosion or a poor connection. Locate any potential problems and correct them. Check each of the disconnected sections of the system for any ground-fault conditions that will need to be investigated and repaired. Power up the system and observe the normal start-up procedure. Check to see that there is proper AC and DC current and voltage where they belong.

Figure 5-32

A hydrometer used to test the electrolyte in a lead-acid battery.

Figure 5-33

Check electrical connections in all junction boxes when performing maintenance.

Case Study

—Torresen Marine, Muskegon, Michigan

Solar Photovoltaic System

At the time of its completion, the solar PV installation at Torresen Marine represented the largest solar-powered project in the state of Michigan. The $740,000 system is the result of a joint venture between Inovateus Solar, a supplier of solar equipment operating in South Bend, Indiana, and Chart House Energy, a renewable independent power producer operating in Chicago, Illinois (Figure 5-34).

The solar photovoltaic system consists of 750 panels, each rated at 200 watts, creating a total output of 150 kW (Figure 5-35). This equates to an estimated output capacity of almost 189,000 kWh per year. At current utility prices in Michigan, this means the system will produce $85,000 of power per year. The energy produced will offset the power of the Torresen Marine facility by 30% and can produce enough electricity to power 20 homes in the Muskegon area.

The panels will be located atop Torresen's existing 28,000-square-foot sailboat storage facility.

Torresen was one of the first marinas in the state of Michigan to achieve a certified clean business standard. In the words of Robert Rafson, president of Chart House Energy: "They recycle waste from boat engines, as well as all of the aluminum and stainless steel waste. The way in which they operate their business is very eco-friendly, making them a perfect candidate for solar power."

Figure 5-34

Installation of solar panels.

Figure 5-35

Electrical control panel for PV modules.

COURTESY OF DONALD STEEBY

UNIT 2
Wind Energy: Setting Sail for a New Power Alternative

INTRODUCTION TO WIND ENERGY

A BRIEF HISTORY OF WIND POWER

Throughout the years, the wind has proven to be a free, clean, and inexhaustible source of energy. The wind has been used for powering sailing ships for many centuries, and many countries have owed their prosperity to their skill in sailing. Since the early American conquest of the Great Plains, the wind has been a useful tool in the United States as well. By harnessing the **kinetic energy** of the wind and turning it into mechanical energy, the early U.S. settlers knew that the wind could be utilized for several different purposes. For instance, windmills were frequently used on early American farms and ranches for drawing water from surface wells (Figure 6-1). The water from these wells was used for watering livestock as well as for irrigation purposes.

Windmills were also used by milling companies for such applications as grinding grain. In fact, the first corn-grinding wind turbine was built in Holland in 1439. By 1600, the most common wind turbine was the tower mill. The word *mill* refers to the process of grinding or milling grain, and this application was so common that most wind turbines were referred to as windmills. The Dutch people of Holland were the first to develop and perfect the windmill. Since the 16th century, they have inhabited the low areas of their European country that were primarily made up of former lake bottoms. Their windmills made habitation of this area possible by draining the water from the lakes and helping to keep their land dry (Figure 6-2).

The Dutch settlers who came to America in the mid-1700s brought this wind technology with them. From the late 1880s until the mid-1930s, there were approximately 6 million mechanical windmills in operation in the United States, providing water from wells and helping to tame the American West. Many of these units are still working satisfactorily today.

The U.S. wind industry got its start using wind turbines to generate electricity in

COURTESY OF DONALD STEEBY

Figure 6-1

An early American windmill.

Figure 6-2

A typical windmill found in Holland.

California during the 1970s as a result of the first oil embargo, which caused a dramatic increase in the price of electricity. At that time, the California wind industry benefited from federal and state incentives as well as state-mandated utility contracts that guaranteed a fair market price for wind power. As a result, 1,500 megawatts of wind turbines were installed in California, supplying approximately 1% of the state's energy demands yet representing nearly 90% of global installations for that time (Figure 6-3).

Figure 6-3

California saw a huge increase in wind energy production in the 1970s.

When federal tax incentives expired in the mid-1980s, the growth of the U.S. wind energy came to an abrupt halt. Europe then took the lead in wind energy, propelled by aggressive renewable energy policies that were enacted between 1974 and 1985. Technological advances resulted in significant increases in wind turbine power as global industries grew well into the 1990s. In 1998, the average wind turbine had an average capacity that was 7 to 10 times greater than that of the turbines of the 1980s, while the price of wind-generated electricity fell by nearly 80%. By 2000, Europe possessed more than 12,000 megawatts of wind power, compared to only 2,500 megawatts in the United States.

Today, however, as a result of many states requiring that electricity producers obtain a percentage of their supply from renewable energy sources, wind energy is experiencing a revival. States such as Texas and Iowa have enacted **Renewable Portfolio Standards** that have created an environment for stable growth in wind technology. A Renewable Portfolio Standard (RPS) provides states with a mechanism to increase renewable energy generation using a cost-effective, market-based approach that is administratively efficient. An RPS requires electric utilities and other retail electric providers to supply a specified minimum amount of customer load with electricity from eligible renewable energy sources. The goal of an RPS is to stimulate market and technology development so that, ultimately, renewable energy will be economically competitive with conventional forms of electric power. By 2005, the United States once again established itself as a world leader in new wind energy after trailing Germany and Spain for the previous decade. This resurgence is attributed to a growing interest in renewable energy, continued improvements in wind technology and its performance, and an increase in supportive policies.

Today's windmills are more commonly referred to as wind turbines because they are predominantly used to generate electricity, in much the same way that a gas or steam turbine generates electricity. Wind turbines constitute the fastest growing source of alternative energy used today (Figure 6-4). In 2006 alone, the total output of all wind turbines being utilized in the United States equaled 9,000 megawatts

Things to Know

RENEWABLE PORTFOLIO STANDARDS

An RPS is a state policy that requires electricity providers to obtain a minimum percentage of their power from renewable energy resources by a certain date. Currently, 24 states plus the District of Columbia have RPS policies in place. Together, these states account for more than half of the electricity sales in the United States.

Figure 6-4

Wind turbines constitute the fastest growing source of alternative energy used today.

COURTESY OF DONALD STEEBY

of generated electrical power. This number equates to 26 billion kilowatt-hours of electricity produced. To put this figure into perspective, the average household consumes approximately 11,000 kilowatt-hours of electricity annually. One megawatt equals 1 million watts of electricity (mega = million). One megawatt of energy can generate about 8.77 kilowatts annually after it is de-rated for efficiency. Therefore, 1 megawatt of energy can generate enough electricity to satisfy approximately 800 average homes on a yearly basis. Wind power is one of the country's largest sources of new power generation.

In 2008 alone, 8,500 megawatts of wind power were installed, providing 42% of all new generating capacity added in the United States. Although these numbers may seem huge, the current amount of electrical power produced from wind turbines in the United States constitutes less than 1% of the total electrical power produced each year.

HOW A WIND TURBINE WORKS

Air is a very fluid substance—that is to say, it flows very easily, just like water. In fact, air is even more fluid than water. This is why wind turbines are engineered to take advantage of the phenomenon of fluid flow that air possesses. Wind is created naturally by the uneven heating of the sun on the earth's surface. So it could be said that wind is actually a form of solar energy.

The heating of the earth's surface causes the air to become warmer. As warmer air rises, cooler air rushes in to fill the space that it leaves behind. This effect is what generates the winds that blow across the earth's surface (Figure 6-5).

Figure 6-5

Wind currents are affected by the surfaces of both land and water.

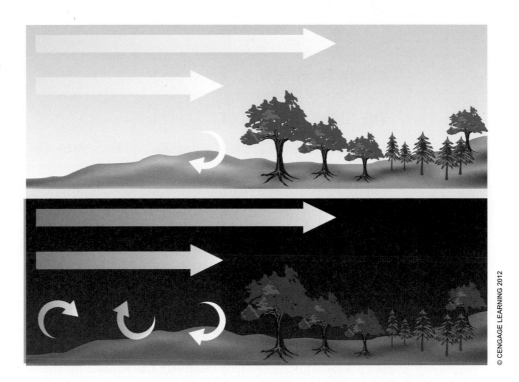

© CENGAGE LEARNING 2012

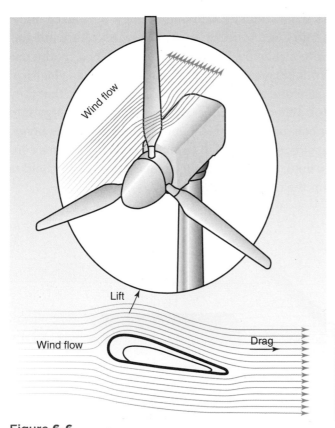

© CENGAGE LEARNING 2012

Figure 6-6

The aerodynamics of a wind turbine.

In addition to this, the difference in the temperature of these air masses creates a difference in the barometric pressure. Air naturally travels from a higher to a lower barometric pressure. As the barometric pressure within a particular area increases, the strength of the wind increases. Obviously, a wind turbine will increase its electrical output and improve its efficiency with an increase in wind speed. Therefore, the proper location that is chosen to install a wind turbine is dependent upon the average annual wind speed and duration for that location. Wind flow patterns are modified by the earth's terrain, bodies of water, and by the amount and type of vegetation.

The blades or propellers on the rotor of a wind turbine operate on two principles: lift and drag. Without getting into a highly sophisticated discussion on aerodynamics at this point, lift will be defined as the force of the wind that acts perpendicular to its flow; and drag is the force of the wind that acts parallel to its flow. These forces are the reason why the turbine's blades utilize an airfoil design. Figure 6-6 shows a cross section of a turbine blade that incorporates this type of

Things to Know

THE BERNOULLI EFFECT

The reason that airplanes can fly and people can sing is based upon Bernoulli's principle. Named after the Dutch-Swiss mathematician Daniel Bernoulli, this principle states that an increase in air speed occurs simultaneously with a decrease in air pressure. The most common example of this is in the action of an airfoil. The shape of an airplane wing is such that air flowing over the top of the wing must travel faster than the air flowing under the wing, and so there is less pressure on the top than on the bottom, resulting in lift.

design. Notice that the blade resembles an airplane's wing in that it is rounded on one surface and tapered to a relatively flat surface on the other side. This airfoil design creates a change in air pressure between the top and bottom of the blade as the air passes over it. The decrease in air pressure over the top or curved side of the blade, and simultaneous increase in air pressure on the bottom side, creates the lift that allows airplanes to fly and wind turbines to become more efficient.

A wind turbine converts the wind's kinetic (moving) energy into mechanical energy that is then converted into electrical energy. The process of a wind turbine converting wind energy into mechanical power is accomplished by the use of a generator. The generator produces electricity that is connected to an electrical panel or to an electric grid. A generator is similar to an electric motor in that they both consist of a rotor and stator. Whereas a motor converts electrical energy into mechanical energy, a generator converts mechanical energy into electrical energy. Voltage is created through magnetic induction. This induction occurs when a conductor cuts through a magnetic field. As the conductor moves through the field, a potential difference is created in the conductor (Figure 6-7).

Figure 6-7

The principle of how magnetic induction can create electricity.

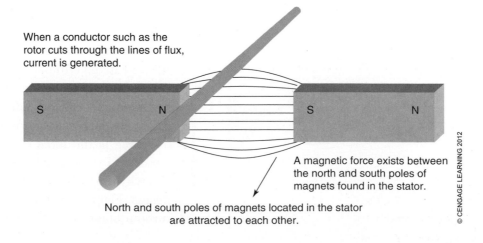

When a conductor such as the rotor cuts through the lines of flux, current is generated.

S   N   S   N

A magnetic force exists between the north and south poles of magnets found in the stator.

North and south poles of magnets located in the stator are attracted to each other.

Figure 6-8

A wind turbine generator consists of a rotor and stator, similar to an electric motor.

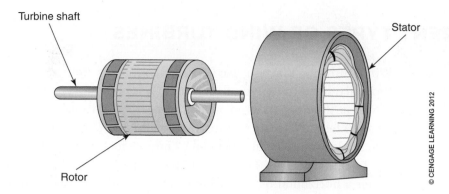

Turbine shaft

Stator

Rotor

© CENGAGE LEARNING 2012

The size of this potential difference is based upon the strength of the magnetic field and the speed of the conductor passing through it. In a generator, the rotor acts as the conductor and the stator as the magnetic field. Essentially, generators are large quantities of copper wire spinning around inside of very large magnets, at very high speeds (Figure 6-8).

As mentioned earlier, today's modern wind turbines are aerodynamically designed to increase their efficiency by capturing the maximum amount of the wind's energy (Figure 6-9). The wind first makes contact with the blades and cause the rotor to turn. The rotor rotates the shaft, which in turn spins the generator. The generator may produce either AC (alternating current) or DC (direct current). If it produces direct current, an inverter is used to convert the direct current to alternating current. The reason for this conversion is two-fold. First, it is much easier to transport alternating current over power lines without losing efficiency, and second, typical households use alternating current as their primary source for electrical usage. The physical connections between the turbine and the home's electrical connection will be explained later in this unit.

Figure 6-9

A cut-away view of a small wind turbine.

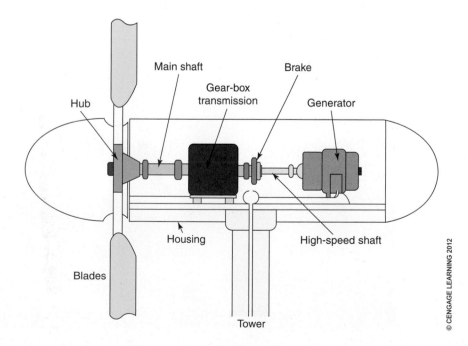

Hub

Main shaft

Gear-box transmission

Brake

Generator

Housing

High-speed shaft

Blades

Tower

© CENGAGE LEARNING 2012

DIFFERENT TYPES OF WIND TURBINES

Wind turbines that are used for residential or light commercial operation are typically sized to deliver electricity at a rate of between 900 and 10,000 watts of power at their tested wind speed. They are usually mounted on a tower that is at least 30 feet tall so the turbine can be situated above any nearby wind obstacles. The blade area is usually between 7 and 25 feet in diameter. Today's turbines that are installed for residential and commercial use are primarily configured with either a horizontal or vertical axis. Horizontal-axis wind turbines are sometimes referred to as propeller-type turbines, whereas vertical-axis turbines are sometimes referred to as eggbeaters. Following is a detailed description of both types.

Horizontal-Axis Wind Turbines

Horizontal-axis wind turbines, or **HAWTs**, are what most people think of when they envision a wind turbine (Figure 6-10).

They are configured with a horizontal drive, with rotor blades that are perpendicular to the ground. Most of today's horizontal-axis wind turbines have three blades. Because of this horizontal configuration, HAWTs receive the wind's power through their whole blade rotation, making them very efficient. What is specific to some of these types of wind turbines is that they require a steering mechanism to align the turbine into the wind when it changes direction. On smaller turbines, this is accomplished by the use of a wind vane mounted at the rear of the horizontal

Figure 6-10

A typical horizontal wind turbine configuration.

Figure 6-11

A typical horizontal wind turbine utilizing a steering mechanism.

axis. This steering mechanism is much like the rudder of a boat—it steers the blades of the rotor into the wind's direction (Figure 6-11).

On larger, commercial-type wind turbines, sophisticated equipment such as an **anemometer** connected to a microprocessor computer is used to automatically steer the wind turbine into the direction of the wind to maximize its efficiency.

Horizontal-axis wind turbines can be classified as upwind or downwind machines (Figure 6-12). Upwind-type turbines have the rotor facing into the wind, similar to a propeller on an airplane (Figure 6-13). The advantage of this upwind position is there is reduced tower shading—interference of the tower with the wind direction. The disadvantage of this configuration is that the rotor must be positioned far enough away from the tower to avoid any problems with a blade strike. Also, the blades must be rigid enough to avoid them being bent back into the tower.

Figure 6-12

An illustration showing the difference between an upwind and downwind horizontal-axis turbine.

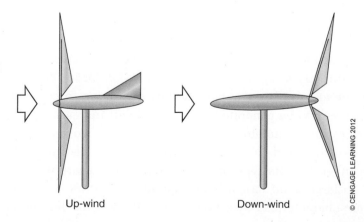

Up-wind   Down-wind

Figure 6-13

An upwind horizontal-axis turbine.

Figure 6-14

A downwind horizontal-axis turbine.

The downwind turbine has its rotor on the back side of the turbine (Figure 6-14). This model is designed to seek the wind; thus there is no need for a separate **yaw** mechanism. The rotor blades can be flexible since there is no danger of a tower strike. The flexing blade has two advantages: they can be less expensive to make and they can relieve stress on the tower during high-wind conditions. This is a result of the hinged design that allows the blade to flex back, dissipating energy for speed control. The flexible-blade advantage can also become a disadvantage, as the flexing may cause blade fatigue. Tower shading is also a problem due to the fact that the rotor blade passes behind the tower.

Horizontal-axis wind turbines can be divided into two main groups depending upon the type of generator they are using: asynchronous (induction) generators, and synchronous generators. **Asynchronous generators** require that they be connected directly to an electrical grid. As wind speed increases, the tendency for the rotor to turn faster is balanced by a reactive force from the grid, allowing more power to be generated for a very small increase in rotational speed. This type of turbine generator is characterized by the slow, almost constant rate of the blades turning and can be fed directly into the electrical grid without the need of an inverter. This type of turbine includes a gearbox and is the type usually seen on large wind farms. **Synchronous generators** are found on wind turbines that are characterized by a variable rate of rotation that increases with the wind speed. These types of turbines have outputs that change frequency and must be first rectified and passed through an inverter before being fed into the electrical grid. The rotational speed of the rotor is much higher than with the asynchronous type, and a gearbox normally is not required.

Vertical-Axis Wind Turbines

Vertical-axis wind turbines, or **VAWTs**, are manufactured with their shafts mounted on a vertical axis that is perpendicular to the ground, hence the name (Figure 6-15).

These turbines are unique in that they are always aligned to the wind, which means there is no auxiliary steering device that is necessary for a change in the wind's direction. One of the advantages of these types of turbines is that they can be installed at a lower elevation than a horizontal-axis wind turbine, and thus they do not require a tall tower. Also, the generator and gearbox can be placed closer to the ground. Because of these advantages, installation costs are usually less and they do not require as large of a footprint for installation. Furthermore, they tend to be quieter than their VAWT counterparts. However, due to their reciprocating rotation, which requires their airfoil surfaces to backtrack against the wind, VAWTs may not have the overall efficiency of horizontal-axis turbines. There are two main variations of vertical-axis wind turbines. One type is the **Darrieus wind turbine** (Figure 6-16). This design uses lift forces generated by airfoils and was patented by the French aeronautical engineer Georges Jean Marie Darrieus in 1931. The Darrieus type of turbine is theoretically just as efficient as the horizontal-axis wind turbine, if the wind speed is constant. However, in reality, the efficiency of this type of vertical-axis wind turbine is rarely realized due to the stresses that are imposed by variations in wind speed. Also, it is difficult to protect the Darrieus turbine from extreme wind conditions.

The second type of VAWT is the Savonius model, which uses drag forces to generate electricity. The Savonius model operates in the same fashion as a cup anemometer. such as the type found in a weather station (Figure 6-17).

The speed of the cups cannot rotate faster than the speed of the wind; therefore, the Savonius-type vertical-axis wind turbine

COURTESY OF DONALD STEEBY

Figure 6-15

A vertical wind turbine configuration.

Figure 6-16

A Darrieus-type wind turbine.

Figure 6-17

Savonius-type wind turbines are similar to a cup-type anemometer.

turns slowly but generates a high torque. This factor does not make it very suitable for generating electricity, as all turbine generators need to be turning at hundreds of RPMs to generate high voltages. A gearbox could be added to the Savonius, but the added resistance would require very strong winds to get the blades spinning.

Chapter

7

CONSIDERATIONS FOR RESIDENTIAL AND LIGHT COMMERCIAL WIND TURBINES

With energy costs continually on the rise, it makes sense for homeowners and owners of commercial buildings to consider harnessing the wind's energy for their residential and commercial electrical usage. Deciding to purchase and install a wind turbine for residential or light commercial use can be a major decision, and there are many turbines currently on the market from which to choose. But before the decision is made to invest in a turbine, several questions need to be answered:

- Which wind turbine model will best suit the needs of the user?
- Is there enough wind to sustain the turbine's capacity?
- Are large towers allowed in the neighborhood or business area?
- What size wind turbine is required?
- How much electricity will need to be produced?
- What about installation and maintenance support?

It is important to consider all of these factors in order to make an informed buying decision regarding the purchase of residential and light commercial wind turbines.

MODEL SELECTION

To get an idea which wind turbine is the best fit for a given building or residence, several factors should be evaluated. These include the following:

- Rotor diameter and sweep area
- Tower top weight
- Cut-in wind speed
- Rated wind speed
- Rated output
- Peak output
- Annual energy output
- RPM at rated output
- Generator type
- Governing system
- Shut-down mechanism
- Controls included
- Warranty

Table **7-1**

Comparison of Data on Three Different Wind Turbines

Manufacturer	Model A	Model B	Model C
Rotor diameter in feet	8	9	12
Sweep area in square feet	50.3	63.6	113.1
Tower Top Weight (pounds)	75	65	175
Cut-in Wind Speeds in mph	8	7	6
Rated wind speed in mph	30	22	25
Rated Output in watts	950	660	1,010
Peak Output in watts	950	710	955
Annual Energy Output in kWh			
kWh per month @ 8 mph wind	600	800	1,800
kWh per month @ 9 mph wind	900	1,100	2,500
kWh per month @ 10 mph wind	1,200	1,500	3,200
kWh per month @ 11 mph wind	1,600	1,800	4,000
kWh per month @ 12 mph wind	2,000	2,200	4,800
kWh per month @ 13 mph wind	2,500	2,600	5,500
kWh per month @ 14 mph wind	2,800	3,000	6,500
RPM at Rated Output	1,200	550	365
Generator Type	Permanent Magnet	DC Generator	Brushless Alternator
Governing System	Tilt-Up Furling	Angle Furling	Blade Pitching
Shut Down Mechanism	Dynamic Brake	Dynamic Brake	Disc Brake
Controls Included	Battery Controller	Battery Controller	Controller & Dump Load
Warranty	2 years	5 years	2 years

© CENGAGE LEARNING 2012

Table 7-1 compares three different hypothetical wind turbines based upon the above-mentioned factors. These data should be made available by the turbine manufacturer in order for the consumer to make an educated buying decision. Refer to this table while each of these factors is discussed.

Rotor Diameter and Sweep Area

Rotor diameter is the length of the rotor from tip to tip of the blades. It is important to note that the turbine's rotor includes the actual blades and hub assembly. This assembly is essentially the wind collector for the turbine because the blades harvest the wind, and in turn convert the wind into electrical energy. The larger the rotor, the greater the amount of electricity produced. In fact, the output of the wind turbine is directly proportional to the sweep area, and this output essentially doubles as the sweep area doubles. The sweep area is the total area that the blades encompass and is calculated in the same manner as the area of a circle. The sweep area can be one of the most important factors to consider when comparing the power output of various turbines. Figure 7-1 compares the output power of the turbine based upon the rotor diameter.

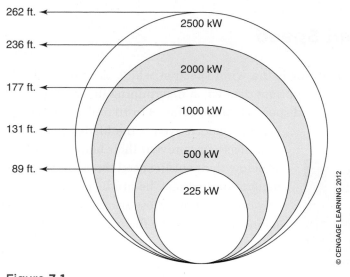

Figure 7-1

Comparison of rotor diameter in relation to power output.

Things to Know

CALCULATING THE AREA OF A CIRCLE

The formula for calculating the area of a circle is $A = \pi r^2$.
 Where:
 A = area
 r = radius
 π(pi) = 3.14
The radius is equal to one-half the diameter. For example, if the diameter of the sweep area is 10 feet, the radius is equal to 5 feet. Therefore, the area is expressed as $5^2 \times 3.14 = 78.5$ square feet.

Tower-Top Turbine Weight

The turbine's tower is usually the most expensive part of the system and can be directly related to the turbine's durability. Usually, a heavier turbine often translates into a more durable system. Remember that the tower is designed for only one height and one size of turbine, so proper tower sizing is critical. The standard rule of the industry is: the lowest point of the turbine's rotor should be at least 30 feet above the nearest object within 300 feet. This rule translates into towers that are typically between 50 and 150 feet tall. It is best to consider these factors and carefully choose the correct tower the first time, as it would be next to impossible to replace it after the turbine is erected.

Cut-in Wind Speed

This factor is the wind speed at which the turbine begins to produce electricity and is important data when determining the average wind speeds at a given location. Wind speeds that are below 8 to 10 mph essentially do not produce any measurable energy, even if the blades are turning. In fact, some turbine manufacturers prevent their blades from spinning anytime the wind is below their cut-in speed threshold. Any small amount of electrical generation that may be produced at these low wind speeds is generally offset by power losses and voltage drops as a result of long wiring connections between the turbine and the grid hook-up.

Green Tip

Selecting a Wind Turbine—Understanding Output Power

When selecting a wind turbine, it is important to understand the following formula:

$$P = 1/2d \times A \times V^3$$

Where:
P = power (expressed in watts)
d = air density
A = sweep area of the rotor
V = wind speed

Notice that in this formula, the wind speed (V) is cubed, otherwise expressed as the wind speed to the third power. In other words, the power output of the turbine (in watts) is directly proportional to the cube of the wind speed. This fact is important to remember when considering the rated wind speed data because as shown by this formula, when the speed of the wind doubles, the output power of the turbine increases by 800%!

Rated Wind Speed

The rated wind speed is equal to the speed of the wind when the turbine has reached its rated output. This is an important piece of data to consider, as illustrated by the following example. Let us compare two wind turbines. Turbine A has an output rating of 1,000 watts at a rated wind speed of 20 mph. Turbine B has an output rating of 1,000 watts at a rated wind speed of 40 mph. Notice that Turbine A can produce the same output at half the wind speed, even though both turbines are producing a peak output of 1,000 watts at 40 mph. Using the formula for calculating the turbine's power output, Turbine B is producing only 125 watts, or 1/8 of the amount of Turbine A at a 20-mph wind speed. Another way of looking at this fact is that if the wind speed is doubled, the power output is increased by 800%—up to the turbine's rated output.

Rated Output

This information is the standard output in watts that the manufacturer designs the wind turbine to provide. It is usually considered a safe level at which the turbine can operate to prevent it from blowing apart. Even though the turbine may have the potential for a higher output, the rated output is generally closer to the governing wind speed. A **governor** is a device that prevents the turbine from spinning out of control in high winds.

It is important to compare the rated output to the rated wind speed data. As mentioned, a higher rated output in wattage at a lower rated wind speed will result in more capacity over various wind conditions. Better yet, compare the kilowatt-hour output per month of electricity that various turbines will produce over various average wind speeds.

Peak Output

The peak output of a given wind turbine may be equal to the rated output, or even higher. Turbines reach their peak output when they are performing above their rated wind speed. This is usually while they are governing to prevent the rotor from being blown apart by high wind speeds. Although the manufacturer may promote the turbine's peak output as a marketing factor, a better indication of overall performance would be to study what the turbine will do at the average wind speed for a given area. This is due to the fact that maximum wind speed usually occurs only for a fraction of the time for a particular area over a given year.

Annual Energy Output

Annual energy output is the approximate amount of power in kilowatts that the average wind turbine will produce per year at a given average wind speed. The information listed here should be used to match the building's electrical requirements and is a good tool when sizing the wind turbine. The performance of a given turbine may vary depending upon the actual average wind speed at the given site. This is why it is important to have accurate data when selecting the best turbine for the best results. Keep in mind that the annual energy output is based upon locations that are from sea level to 1,000 feet in elevation and must be adjusted for higher altitudes due to lower air density. As stated earlier, the average home consumes between 800 to 1,000 kilowatt-hours of electricity on a monthly basis. This is in contrast to a very energy-efficient home or cottage that may use only 100 to 500 kilowatt-hours per month. In comparison, a small business or farm may use upward of 1,500 to 2,500 kWh for the same period. It may be best to consult the manufacturer or dealer to help sort out all of this data when making a buying decision.

RPM at Rated Output

The RPM, or blade revolution speed, at the rated power output is related to two factors: the amount of noise created by the rotor and the overall durability of the turbine. Usually, a smaller rotor will mean that the blades will spin faster and increase the noise level, as well as contribute to wear and tear on the bearings. Conversely, a slower rotor speed will equate to less wear on moving parts and ultimately will mean a turbine that will last longer and perform more quietly. This doesn't necessarily mean that a lower RPM level results in lower power output. Neither does it mean that a higher RPM will produce more. Usually, the turbine's alternator is matched to the rotor speed to get the maximum amount of power out of the available wind.

Generator Type

The three main types of generators in a wind turbine are: direct-current generators, permanent magnet (PM) alternators, and brushless alternators. The generator is the device located in the wind turbine's housing that is used to convert the wind's energy into electrical energy using the principle of **electromagnetic** induction, the same principle governing how an electric motor works. In general, PM alternators use an electromagnet in the rotor that has many strong magnetic poles. This is done so that the generator can produce a higher output voltage at a lower RPM. The disadvantage of using this method is that it is more difficult to regulate over a larger RPM range. PM alternators are lighter weight, less expensive to manufacturer, and are less complicated than other types of generators.

Brushless alternating-current alternators generate electricity the same way that direct-current generators do, by generating an electrical current as the rotor turns within a stationary coil, known as the stator. With a brushless alternator design, permanent magnets are bonded to the rotor, which takes the place of brushes. This results in a more efficient design and less wear on the generator.

Governing System

By definition, a governor is a device used to maintain a uniform speed regardless of changes in load. The governing device on a wind generator is necessary for two reasons. First, it protects the turbine's generator from overproducing power and eventually burning itself out. Second, it protects the turbine from spinning out of control and blowing apart in high winds. There are two main types of turbine governing systems. The first is accomplished by reducing the area of the rotor facing the wind, known as furling. The other type is by changing the pitch of the blades.

The blade-furling system of governing a turbine is accomplished by tilting or furling the rotor directly up, on its side, or at an angle (Figure 7-2). A blade pivot point causes the rotor to pitch away from the wind direction due to the high pressure exerted on the blades. Although very reliable, this type of governing system does reduce system output at excessive wind speeds due to the fact that the turbine is no longer oriented into the wind.

The pitched-blade method of governing is accomplished by automatically pitching the blades out of their optimum aerodynamic angle during high winds. This type of governing system is usually found on large-scale commercial wind turbines (Figure 7-3).

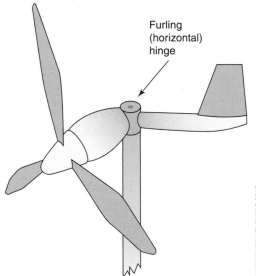

Furling
(horizontal)
hinge

© CENGAGE LEARNING 2012

Figure 7-2

A wind turbine with a horizontal furling hinge.

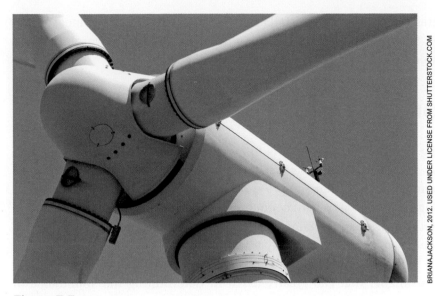

BRIAN A JACKSON, 2012. USED UNDER LICENSE FROM SHUTTERSTOCK.COM

Figure 7-3

The pitched-blade type of governing system is usually found on large-scale commercial wind turbine

Blade-activated governors have more moving parts than furling systems and are more complex. More moving parts equates to more maintenance and more potential breakage. The upside to pitched-blade governors is that they produce a higher level of output power at higher wind speeds as compared to the furling system.

Shut-Down Mechanism

In excessive winds, during routine maintenance, or when the turbine is simply not needed, it may be necessary to stop it completely. This can be accomplished by use of the shut-down mechanism. This device typically shuts down the turbine either by dynamic braking or by the use of mechanical brakes. Dynamic braking is unique to permanent magnet alternators. It is done by electrically shorting out the magnets of the alternator, thereby overpowering the ability of the rotor to spin and causing it to come to a complete stop. Mechanical or disc braking systems for the wind turbine are similar to the brakes used on an automobile. They may utilize a hydraulic system and are usually more reliable than the dynamic braking system.

Controls

The controls for the turbine include a rectifier, brake, dump load, and metering. These items should be standard equipment with the purchase of the wind turbine. Do not forget to include these items in the overall costs if they are optional equipment. In addition to controls, such things as batteries, wiring connections, crane and rigging costs, shipping charges, and, of course, the tower need to be included when estimating the overall cost of the project.

Warranty

The warranty reflects the manufacturer's confidence in the product. It usually covers replacement of parts due to defects in materials and workmanship. It does not always cover the labor to replace these parts. Furthermore, a warranty does not include failures due to improper installation, lack of routine maintenance, abuse, or acts of God. Sometimes the manufacturer will offer an extended warranty for an additional price.

WIND QUANTITY CONSIDERATIONS

The qualifying question that first needs to be answered is: Will there be enough wind in the area where the turbine is to be installed? In other words, does the wind blow strong enough for a sustained period of time in order to justify the economics of installing a residential or light commercial wind turbine? The answer to this question is not always easily obtained. Wind velocities and their abundance can vary greatly between areas that are only a few miles apart due to the differences in terrain and the amount of buildings in the path of the wind. However, there are resources available that can greatly assist in determining the abundance of wind in a given area. These resources include wind maps such as the one shown in Figure 7-4. The highest average wind speeds in the United States are mostly found along such areas as seacoasts, ridgelines, or plateaus, and along the Great Plains region. This does not mean that other areas are not suitable for small wind turbine use. If the local terrain features a hilltop, ridge crest, or vast exposed area, a residential or light commercial wind turbine could still be economically feasible. More detailed wind resource information can be obtained from the National Wind

Figure 7-4

Wind strength in various areas of the United States.

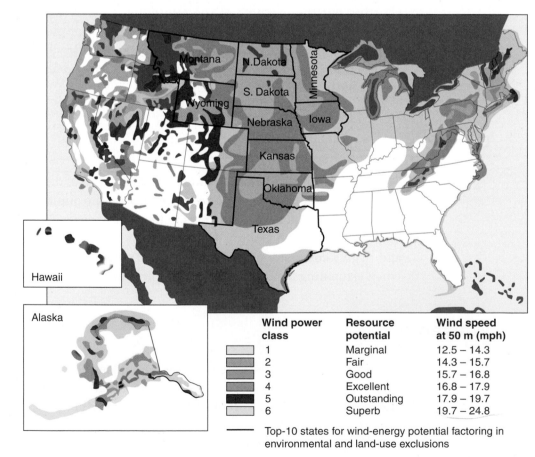

Wind power class	Resource potential	Wind speed at 50 m (mph)
1	Marginal	12.5 – 14.3
2	Fair	14.3 – 15.7
3	Good	15.7 – 16.8
4	Excellent	16.8 – 17.9
5	Outstanding	17.9 – 19.7
6	Superb	19.7 – 24.8

— Top-10 states for wind-energy potential factoring in environmental and land-use exclusions

Technology Center website, from the U.S. Department of Energy Wind Technology Center website, or from the local meteorological center located at the nearest municipal airport. If one chooses to use wind resource information from a nearby airport, keep in mind that local terrain influences and other factors may cause the wind speed to be different at the airport's location than from another given location.

Another useful measurement of the abundance of wind in a particular area can be obtained by observing the growth of the local vegetation, especially evergreen trees. These trees can be permanently deformed by being subjected to strong sustained winds over a long period of time. This phenomenon is known as **flagging** and can be used to estimate the average wind speed for a specific area. Figure 7-5 shows the effects of flagging on area vegetation.

If more specific information is required to determine accurate average wind speed for a specific area, wind measurement systems are available. The average cost of this type of system ranges between $600 and $1,200.

Particular consideration must be made when using measurement equipment to ensure that the wind is measured at a high enough altitude to avoid turbulence created by buildings, trees, and other obstructions, as shown in Figure 7-6.

Because air is very fluid, any obstructions in the wind's path of the turbine can create turbulence, similar to the wake behind a boat. The most useful readings would be taken at the same height as the potential turbine's location. Normally this

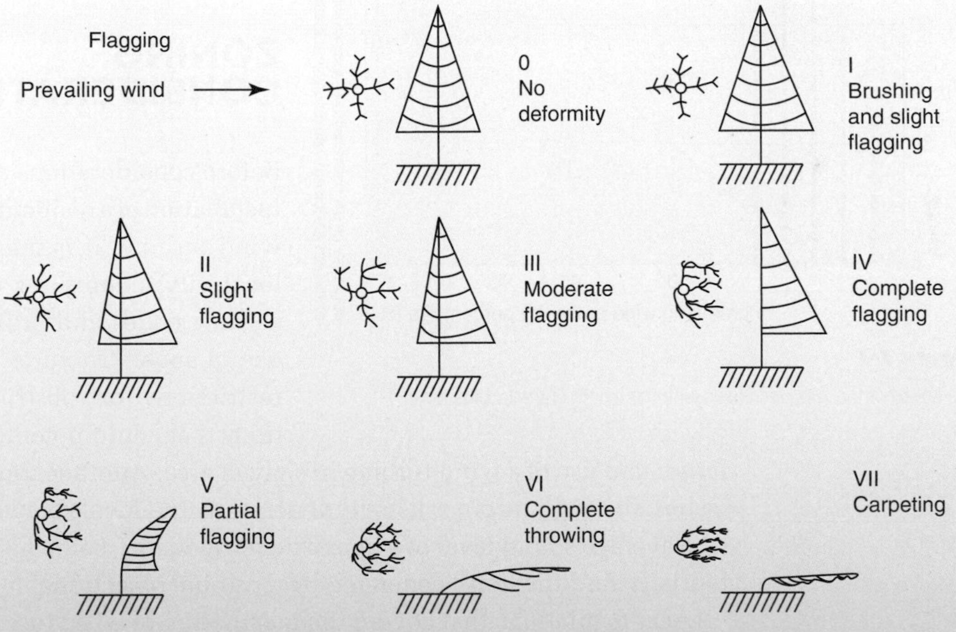

Griggs-Putnam index of deformity

Index	I	II	III	IV	V	VI	VII
Wind mph	7–9	9–11	11–13	13–16	15–18	16–21	22+
Speed m/s	3–4	4–5	5–6	6–7	7–8	8–9	10

Figure 7-5

Flagging is the effect of wind on vegetation and can help determine average sustained wind speeds.

COURTESY OF U.S. DEPARTMENT OF ENERGY, OFFICE OF ENERGY EFFICIENCY AND RENEWABLE ENERGY

Figure 7-6

Less turbulence is encountered if the turbine is placed farther away from obstacles.

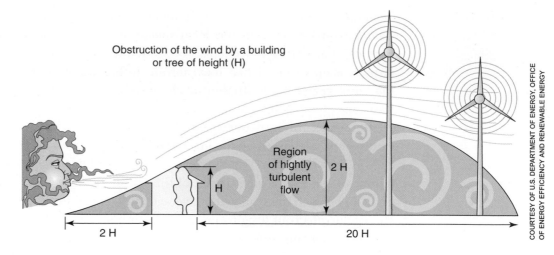

Obstruction of the wind by a building or tree of height (H)

Region of hightly turbulent flow

2 H

H

2 H

20 H

COURTESY OF U.S. DEPARTMENT OF ENERGY, OFFICE OF ENERGY EFFICIENCY AND RENEWABLE ENERGY

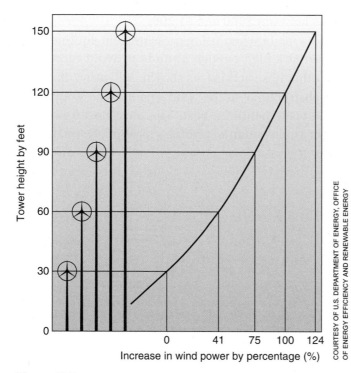

Tower height by feet

Increase in wind power by percentage (%)

COURTESY OF U.S. DEPARTMENT OF ENERGY, OFFICE OF ENERGY EFFICIENCY AND RENEWABLE ENERGY

Figure 7-7

The speed of the wind increases with the height of the turbine tower.

height would be 30 feet above any obstruction within 300 feet of the turbine's tower. Figure 7-7 shows how the tower height is proportional to an increase in wind speed. Notice that by doubling the height of the tower, wind power increases approximately 50%.

Last, one could receive accurate wind information from another site in a given area where there is a wind turbine currently installed.

ZONING CONSIDERATIONS

Before considerations are made regarding the installation of a residential or light commercial wind turbine, it is mandatory to check with local city, county, or township officials regarding codes and ordinances regulating the use of such structures. There may be height restrictions or obstruction-of-view regulations that could potentially impact the installation and use of a wind turbine in a given area. Another consideration is the noise issue. Although turbines usually generate sound levels between 52 and 55 decibels (the average sound level of a domestic refrigerator), some neighborhoods could potentially find the noise generated by a turbine objectionable.

Local regulations that govern the installation of wind turbines can affect mandatory zoning and building permits. For instance, there is usually a required setback distance from the edge of the property that can influence the tower's placement. This setback can also be dependent upon the actual height of the tower as well as the blade length. In addition, the "fall zone" affects the placement of the tower. This is the area around the tower that must be kept free of objects that could be damaged should the tower ever collapse, such as in a severe wind. Obviously, this issue becomes more difficult to address in an urban setting versus one that is more rural. If the wind turbine is to be constructed within a neighborhood that is owned by a condominium association, the area of placement would more than likely be the condominium's common area.

Another local ordinance to consider is one that addresses "shadow flicker." This incident occurs when the tower's moving blades come between the viewer and a light source, such as the sun. When the sun's angle is within the exact path of the moving blades, shadow flicker can cause disruptive shadows on a nearby building, such as a neighbor's house. Although this phenomenon is not an issue for most people, the wind turbine owner must be aware that local ordinances typically require that a neighbor's right to "quiet enjoyment" not be encumbered.

In some cases, the local zoning authorities may require that the wind turbine owner be bonded for any additional expenses. This is usually addressed as a performance bond and can cost as much as 5% of the total project. These additional expenses are to ensure that the wind turbine be:

1. Properly maintained
2. Properly decommissioned

Decommissioning occurs if and when the tower is ever to be deconstructed. The additional money required is typically set up in an escrow account to ensure that if the owner does not maintain the wind turbine, the municipality will.

Green Tip

Wind Turbine Noise Issues

When the decision is made to purchase and erect a wind turbine in the neighborhood, there are usually questions that arise regarding the amount of noise that it generates. When the prospective owner is seeking approval from the local township for a building permit to erect the turbine, the permit is sometimes delayed due to noise concerns by some of the neighbors.

Most well-designed turbines are relatively quiet. Usually if they are generating noise, it is at a time when the wind is creating its own noise—either from trees rustling or buildings rattling. So, what is a reasonable noise level?

(Continued)

Things to Know

FOWL KILL RATES

Some environmentalist groups are concerned that wind turbines are detrimental to the life and well-being of local birds. The concern is that birds will be killed by flying into the spinning blades of a wind turbine. In reality, the greatest mortality risk among birds in the United States is from flying into closed windows and into glass buildings.

Green Tip (Contd)

Consider the fact that the average background noise in a home is about 50 decibels. Compare this to a car driving down the street (60 decibels), or a vacuum cleaner (70 decibels). Now consider that at 50 feet away, a small wind turbine generates a sound level of between 55 and 60 decibels in winds between 15 to 35 mph. Is this an objectionable level? That depends upon the sensitivity of the person who thinks it is objectionable, or maybe the individual just thinks that the look of a turbine in the neighborhood is objectionable. Sometimes the realistic noise level lies in the ear of the beholder.

Green Tip

Calculating the Payback on a Wind Turbine

Here is an example of the cost analysis of a small residential or commercial wind turbine:

CORRECT SIZING

The primary factors that impact the sizing of a wind turbine include the amount of electricity consumed by the residence or business on a monthly basis and whether the homeowner is seeking to supplement the existing electrical service or go completely off the grid. The average home consumes between 800 and 1,000 kilowatt-hours of electricity on a monthly basis. This equates to approximately 10,000 kWh per year. Based upon the average wind speed in a given area, a wind turbine that is rated between 5- and 15-kW output could make a significant impact on the given electrical demand. Another example would be a home requiring 300 kWh of electricity per month located in an area with an average wind speed of 14 miles per hour. This particular application would require a wind turbine with an output of at least 1.5 kW. Most manufacturers and dealer representatives can provide the correct sizing as well as the average annual energy output of a given turbine based upon the average wind speed.

CONNECTING TO THE LOCAL UTILITY GRID

Once the correct wind turbine has been selected for the appropriate site and application, it is time to consider making the proper connections to the building and local utility grid. This can be accomplished by several methods, depending upon how simple or complicated one chooses to make it. Following is a list of several configurations for grid connections based upon the end-user's needs:

* Basic grid-connected wind turbine
* Grid-connected wind turbine with battery backup
* Connecting a wind turbine and backup generator

Basic Grid Connection

This is the simplest way to connect a wind turbine directly to the grid without battery backup. The basic elements are the turbine itself and an electrical inverter (Figure 7-8).

An inverter is a device that converts direct current to alternating current (see the "Green Tip" on Electricity Review). It is essential to understand that the electricity generated from the wind turbine, be it alternating current or direct current, may not be generated at the same voltage, frequency, or current that the grid requires. This is because these elements vary with the RPM of the turbine's rotor. Therefore, an inverter is required to "clean up" the electricity generated by the wind turbine so that it is suitable for

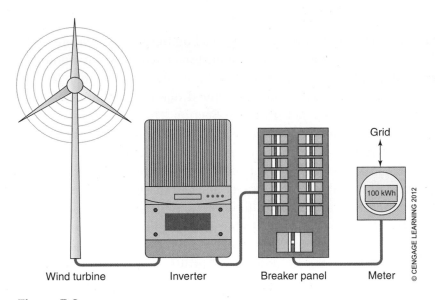

Grid

100 kWh

© CENGAGE LEARNING 2012

Wind turbine   Inverter   Breaker panel   Meter

Figure 7-8

A basic grid connection using an inverter.

connecting to the grid. Once the electricity passes through the inverter, it connects to the breaker panel, where it is used to provide electricity to the home or building.

On days with very little wind, the grid will be the primary source of electricity for the home or building. If the electricity generated by the turbine exceeds the building's requirements, such as on very windy days, the balance passes through the meter and onto the local grid. In this instance, the meter would essentially spin backward, which in some cases could generate income for the homeowner or building owner. As explained in Chapter 5, this setup is referred to as net metering and is a process where the building owner receives full value for the power that is produced by the wind turbine system. By the use of net metering, a home or business can offset the cost of its electric bill with any excess electricity that is produced. When the wind turbine system produces electricity, the power is first used to meet any electrical needs that the building requires. When excess electricity is produced, the power is fed back onto the main utility grid. This effect causes the electric utility meter to run backward, allowing the customer to

Green Tip (Contd)

Annual electrical usage of the building = 10,000 kWh (kilowatt-hours)
Cost of electricity per kWh = $0.13
Average wind speed of the given area = 12 mph
It has been decided to purchase a turbine package that will generate on average 12,750 kWh of energy output at a sustained wind of 12 mph.
Cost for the turbine package = $12,500.00
Using these parameters:
First, calculate the annual electrical cost: 10,000 kWh (annual usage) × $0.13 (cost per kWh) = $1,300.00
Then divide $12,500 (cost of the turbine package) by $1,300 (annual electrical cost) = 9.615 years
Payback = A little under 10 years, if the cost of electricity stays constant (this number will decrease if the cost of electricity increases).

(This assumes that the turbine's generator completely offsets the grid power. Most turbines deliver less power than is expected; therefore, two to three are needed to deliver enough power to sustain a residence off the grid.)

receive credit for the additional energy that is placed on the grid. At the end of the billing period, the utility company credits the customer with the net amount of kilowatt-hours produced, unless the customer uses more electricity than the turbine generates, in which case the customer pays the difference. Under federal law (Public Utility Regulatory Policy Act [PURPA], Section 210), local utilities must allow independent homeowners and businesses to interconnect with the utility grid, and the utility company must purchase any excess electricity that is generated. If the homeowner or business owner is in an area where net metering is not utilized, the utility will install a separate meter and purchase the excess electricity at a wholesale price. The sale price of the excess power produced is usually much lower than the retail price. In some states, excess power credits are carried over to the next billing period for up to 1 year. It is important that the homeowner or building owner check with the local electrical utility company to find out its policy regarding excess electricity that is placed back onto the grid. As mentioned earlier, net metering is currently being offered in more than 35 states.

Grid-Connected Wind Turbine with Battery Backup

Obviously, the first basic grid configuration mentioned earlier does not meet the needs of the building owner under every circumstance. One such case would be if the main electrical service from the utility company were to fail and there was not enough wind to sustain the building's electrical demand. In this situation, the owner may want to consider utilizing battery backup. Under this configuration, a charge controller is placed between the wind turbine and the batteries/inverter, as shown in Figure 7-9. The charge controller serves two purposes. First, it converts the unregulated electricity from the wind turbine into useable storage energy for the batteries. Second, it protects the batteries from situations such as excessive current and overcharging. If the batteries are fully charged, power that is produced by the turbine is directed toward the inverter. The inverter that is used in this configuration is more sophisticated than the one used in the basic configuration (no battery backup) because it needs to be able to utilize power from either the turbine or the battery backup. Furthermore, this inverter must be capable of charging the batteries from grid power if necessary. As in the basic configuration, excess electricity that is generated by the turbine is placed back onto the grid, causing the meter to rotate backward.

Some additional components are essential with this configuration. A transfer switch is included with the inverter, which diverts power from the main breaker panel and meter to a secondary breaker panel that powers only essential loads when the grid goes down. This transfer switch is also essential to prevent any electricity generated from alternative energy from being placed on the grid when it is down.

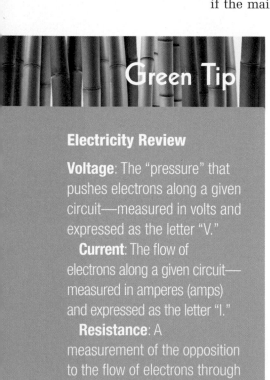

Green Tip

Electricity Review

Voltage: The "pressure" that pushes electrons along a given circuit—measured in volts and expressed as the letter "V."

Current: The flow of electrons along a given circuit—measured in amperes (amps) and expressed as the letter "I."

Resistance: A measurement of the opposition to the flow of electrons through a given circuit—measured

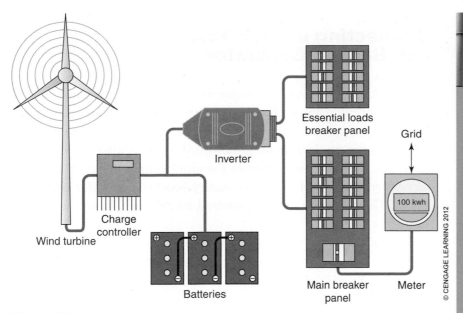

Figure 7-9

A grid connection using battery backup.

This prevents any utility workers from being electrocuted while they are working on the main grid when the power is out. Another feature of the transfer switch is the speed in which it transfers electricity from the main breaker panel to the secondary panel. This switching happens so fast that it is undetectable to the homeowner or building owner. Power under this scenario is now provided strictly by the wind turbine and battery backup. When electricity from the grid is restored, the transfer switch changes back to its original position and normal power is resumed onto the main breaker panel.

Normally, the secondary breaker panel consists of essential loads such as the furnace, lighting, and well pump. It excludes such loads as large electrical appliances (oven, clothes dryer, air conditioning, etc.), as these would be too large for the turbine to accommodate. In addition, sizing the wind turbine and battery pack for these larger loads would be quite impractical. Although this configuration requires a more complicated equipment setup, it is advantageous in that it can provide continuous power to essential household loads, even when the main grid is inoperable.

Green Tip (Contd)

in ohms and expressed as the letter "R."

AC (alternating current): The flow of electrons changes (reverses) direction within a circuit at regular intervals, usually 60 times (hertz) per second.

DC (direct current): Electrons flow in only one direction through a circuit, usually from the source to ground.

An electrical circuit incorporates three basic items: voltage, current, and resistance. These three items are utilized in Ohm's Law, which states that current flowing through a conductor is directly proportional to the potential voltage and inversely proportional to the resistance between them. Ohm's Law is mathematically expressed as $I = V/R$ (where I is amperage, V is voltage, and R is resistance).

Electricity is produced through a generator, which consists of a *stator*, the stationary portion, and the *rotor*, the rotating portion that creates the electromagnetic field. The generator is much like an electric motor that generates electricity through an *electromotive force*.

Protecting Utility Workers

No matter what type of wind turbine system is installed, it must have a means of being disconnected from the grid source in the event of a power outage from the utility supplier. Failure to do so can be potentially dangerous to utility workers who may be repairing the lines, because of the potential to backfeed electricity onto the main grid that is being generated by the wind turbine.

Connecting a Wind Turbine and Backup Generator

This third configuration incorporates a backup generator for use when the turbine and battery backup are not sufficient (Figure 7-10). The main purpose of the generator is to keep the battery pack charged during long periods of main-grid failure, such as after an ice storm. However, it can be configured as the chief source of power if necessary when connected to the secondary breaker panel (for necessary loads). The generator may be powered by gasoline, diesel fuel, natural gas, or propane.

This configuration can utilize a smaller battery pack because of the implementation of the generator, therefore reducing the cost of capital outlay. Keep in mind that peripheral devices such as the proper disconnect switches, fuses, and power-surge protection, as well as a

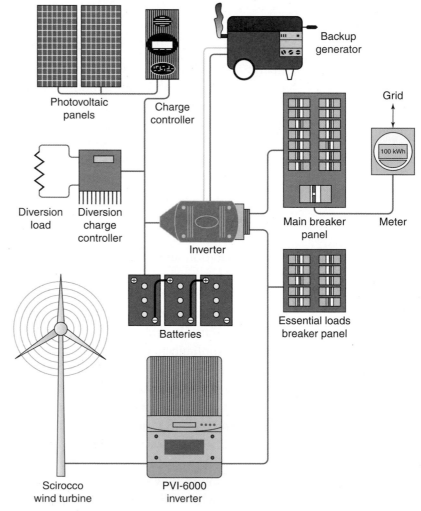

Figure 7-10

A grid connection incorporating battery backup and a backup generator.

transfer switch, will be necessary when utilizing this configuration. Other variations of this configuration are available and may even incorporate the use of solar panels to create a hybrid system.

Following the National Electric Code

Any electrical wiring performed on a wind turbine system must comply with the 2011 National Electric Code (NEC). Much of the information regarding the wiring of photovoltaic systems that was covered in Chapter 5 under "Photovoltaic System Wiring" is consistent with the wiring of wind turbine systems. Article 694 of the 2011 NEC covers the proper wiring installation of small wind electric systems. Close attention must be paid to NEC 694.7, which states that these systems shall be installed only by qualified persons. In addition, Article 694 address the proper installation and configuration of both stand-alone and interactive small wind systems. Other important areas of Article 694 address proper circuit sizing, over-current protection, disconnecting means, wiring methods, proper grounding, the proper use of storage batteries, and connecting to other sources.

SAFETY REVIEW

It is important to remember and follow these safety items whenever making connections to the local electrical grid and to the main/auxiliary breaker panels of any building:

1. Electrical work should always be performed by a qualified, licensed electrician.
2. Always abide by all national and local electrical codes.
3. Always contact the local electrical provider or utility before making any final connections to the grid.

INSTALLATION AND MAINTENANCE ISSUES

Most manufacturer representatives and dealers should have the ability and experience to assist in the installation of the wind turbine that is chosen for a particular residence or for a commercial building. However, some homeowners or building owners may choose to install the turbine themselves. Before attempting to install a typical wind turbine, the following issues should be addressed:

- The impact and effect on local zoning
- The installation of a proper concrete foundation
- The clear understanding of the difference between AC and DC electrical current
- Proper safety practices when installing electrical wiring

- The safe installation and rigging of a 30- to 150-ft. tower
- A clear space for free-fall of the tower
- Routine maintenance

Zoning issues and the proper safety procedures when dealing with electricity have been discussed previously in this chapter. Installation of a properly sized concrete foundation should be reviewed with the dealer or distributor of the wind turbine, and the work should be performed by a qualified concrete contractor.

Installing the Tower

The tower is one of the most important components of a wind turbine. It can constitute over half of the overall cost of the system. Therefore, it is important that the proper installation procedures be utilized. Most small commercial wind turbines are either installed using **lattice towers** with guy wires or are considered freestanding towers (Figure 7-11). Figure 7-12 shows an example of a lattice tower.

Typically, lattice towers or towers made from heavy-duty pipe or tubing require guy wires as a means of support to enhance the stability of the tower, especially in high winds. A guy wire is a tensioned cable that is fastened to the top of the tower and secured to a sturdy base, such as a concrete slab (Figure 7-13).

In some cases, it is advantageous to rig the tower so that it can be lowered for maintenance work or in the event of extremely high winds. This type of configuration is known as a tilt-up tower and incorporates a hinged base for raising and

Figure 7-11

Most small commercial wind turbines are installed using lattice-type towers and guy wires, or are free-standing towers.

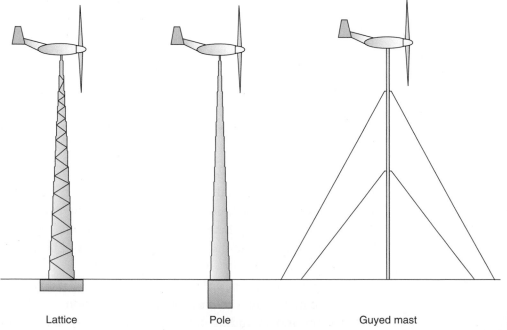

Lattice   Pole   Guyed mast

© CENGAGE LEARNING 2012

Figure 7-12

A lattice-type tower used for supporting a wind turbine.

© ISTOCKPHOTO/TOM GRUNDY

Figure 7-13

Small wind turbines use guy wires to support the towers.

2011 FOTOSEARCH

Figure 7-14

A tilt-up tower can be lowered for maintenance, or to protect the tower in the event of high winds.

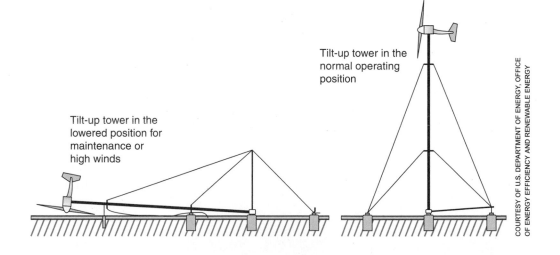

Tilt-up tower in the lowered position for maintenance or high winds

Tilt-up tower in the normal operating position

COURTESY OF U.S. DEPARTMENT OF ENERGY, OFFICE OF ENERGY EFFICIENCY AND RENEWABLE ENERGY

lowering the tower (Figure 7-14). Figure 7-15 shows how this type of tower can be initially erected using a **gin pole** for leverage. A gin pole is a rigid pole that is connected to the base of the tower and uses a pulley on its end for lifting purposes.

Free-standing towers require significantly stronger foundations, as they do not incorporate any guy wires. The main advantage of a free-standing tower is that it utilizes a very small footprint compared to the lattice-type tower. In some cases, a free-standing tower may be as simple as utilizing a common utility pole to mount the turbine onto (Figure 7-16).

Figure 7-15

Erecting a tilt-up tower using a gin pole and lifting point. This point can be attached to a winch, truck, or tractor.

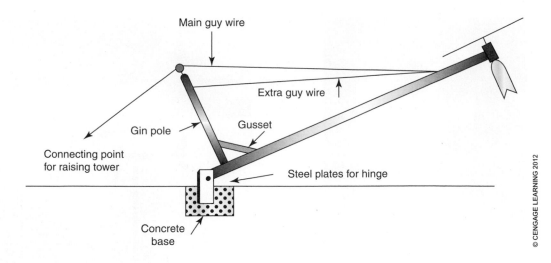

Main guy wire

Extra guy wire

Gin pole

Gusset

Connecting point for raising tower

Steel plates for hinge

Concrete base

© CENGAGE LEARNING 2012

COURTESY OF DONALD STEEBY

Figure 7-16

A common utility pole can sometimes function as a turbine tower.

Performing Routine Maintenance

Every piece of equipment that contains moving parts will require some sort of routine maintenance on a regular basis. Furthermore, moving parts will wear out and eventually fail. Wind turbines are no exception. The extent of the maintenance required on a given wind turbine can vary between manufacturers. Some specifications may require only a visual inspection on an annual basis. Others may get much more involved. Regardless of the manufacturer's recommendations, there are several routine maintenance items that every owner of a residential/light commercial wind turbine should perform.

The first thing to consider regarding routine maintenance is a visual inspection. Are there any signs of extreme wear and tear? Are the main components in sound working order? These are some of the main items to examine first. Second, observe the condition of the blades. Are they in need of repainting? Do the leading edges need to be taped? Remember that the blade material can have an effect on blade durability and life expectancy.

Next observe the condition of the bearings. Do they require grease, or are they permanently sealed? The manufacturer should be consulted regarding bearing lubrication due to the fact that overgreasing can shorten the life of certain bearings. Next, if there is a gearbox, it may need an oil change. Again, consult the manufacturer regarding this item. Any preventative maintenance program should include checking and tightening any nuts and bolts on the wind turbine. This procedure may require the use of a torque wrench so as to not over-tighten certain bolts on the turbine. Normally, most routine maintenance should occur twice per year—once before winter and again before summer. Pick a day when the weather is nice and not too windy. Remember that it is best to purchase a quality piece of equipment. A heavy-duty machine will be more durable and last longer than one made of lighter-weight material.

Field Tip

Maintenance on Tall Towers

A tilt-up tower may be the best choice for a wind turbine when routine maintenance is an issue—especially if one is afraid of heights!

Chapter	# INTRODUCTION TO LARGE-SCALE WIND TURBINES

8

Wind turbines are generally classified into three different categories according to size. Small wind turbines usually are considered to have an output capacity of less than 10 kilowatts. As discussed earlier, these smaller turbines are generally used in residential and light commercial applications. Medium-sized turbines fall into the 10- to 250-kilowatt category and are used for powering such structures as farms, schools, businesses, and other municipal buildings (Figure 8-1).

These mid-sized turbines are usually installed on a stand-alone basis, or may be grouped for use in powering a small town or village. They also may be used in conjunction with other types of alternative energy sources, such as solar panels, to create a hybrid system.

Large turbine applications are sized between 250 kW and 5 megawatts. These behemoths can stand as tall as 250 feet into the air and are used on central station wind farms, and for distributed power (Figure 8-2).

Figure 8-1

Several medium-sized wind turbines.

COURTESY OF ERIN STEEBY

Figure 8-2

A large-scale wind turbine.

COURTESY OF DONALD STEEBY

This section will concentrate on large-scale wind turbines and will cover the following topics:

- The anatomy of a large-scale wind turbine
- Integral controls
- Proper site selection
- Construction and installation
- Electrical connections
- Safety
- Maintenance and repair

THE ANATOMY OF A LARGE-SCALE WIND TURBINE

A wind turbine is made up of essentially the same components, regardless of its size (Figure 8-3). The turbine's blades are similar to those on an airplane's propeller. The blades are connected to a hub, and together the blades and hub make up the turbine's rotor. The rotor is then connected to a main shaft, which is sometimes referred to as the low-speed shaft due to the fact that it spins at the same RPM as the rotor, which usually is not very fast. A gearbox converts the low RPM of the low-speed shaft to a higher RPM, which is in turn connected to the high-speed shaft. A mechanical braking system is connected at the output of the gearbox on

Figure 8-3

Cut-away view of a large-scale wind turbine.

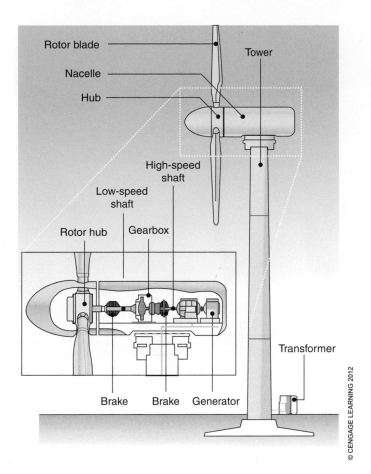

© CENGAGE LEARNING 2012

the high-speed shaft. This brake is very similar to the disc brake on an automobile and is necessary in the event that the rotor needs to be slowed down or stopped. The braking system is considered a fail-safe system in that hydraulic oil pressure is used to hold the calipers off the disc to prevent the unit from braking. In other words, the hydraulic system is holding the brake open. On a command from the central controller, the oil pressure is relieved and a powerful spring presses the brake blocks against the brake disc, slowing down or stopping the rotor.

Things to Know

WIND VANES

Each turbine has a wind vane located on top of the nacelle. The wind vane indicates to the turbine which direction the wind is coming from. Then the rotor and the nacelle can rotate into the face of the oncoming wind.

COURTESY OF JERRY STEEBY

Figure 8-4

The components of the wind turbine are housed in a protective enclosure called a nacelle.

The high-speed shaft is then connected to the turbine's electrical generator. As noted earlier, the generator may produce AC or DC electricity, which is usually filtered, conditioned, and connected to a step-up transformer before it is passed on to the substation and ultimately to the central electrical grid. The rotor, shaft, and gearbox are sometimes referred to as the drive train. The purpose of the gearbox is to increase the RPM of the shaft that is connected to the generator. All of these components are housed in a protective enclosure called a **nacelle** (Figure 8-4).

INTEGRAL CONTROLS

At the heart of today's large modern turbine is the control system (Figure 8-5).

This microprocessor-based controller is involved in most of the decision-making processes and safety features of the wind turbine. This includes monitoring the turbine's normal operational

© ISTOCKPHOTO/ BART COENDERS

Figure 8-5

A modern control panel for a large-scale wind turbine.

mode and also trending measurements for statistical analysis. Controllers can be located in the nacelle or at the base of the turbine.

In a modern wind farm, microprocessor-based controllers are networked together by means of fiber-optic cabling to a central station. Here at the central station, wind technicians can monitor a large number of turbines from a single computerized workstation. Following is a sample of the various parameter points that are monitored and utilized during the central controller's decision-making processes.

Analog Points

Analog points are where measurements produce readings of varying inputs, and include the following:

- Voltage and current readings on all three phases of the generator. (Remember that most large-scale turbines produce three-phase voltage as opposed to smaller turbines that produce single-phase voltage.)
- Temperature inside the nacelle: Maintaining the proper temperature is critical to the microprocessor controls and also to the gear oil in the gearbox and to the bearing grease. Too cold of a temperature can result in the lubricants failing to flow because of excessive viscosity. Too much heat can result in a breakdown of lubricants. Both can contribute to premature gearbox and bearing failure.
- Generator, gear oil, and gear-bearing temperatures: Again, extreme temperatures within these critical components can result in their failure.
- Wind speed and direction of yawing: The turbine's yaw mechanism is used to turn the nacelle into the wind.
- Low- and high-speed shaft rotational speed.
- Variable blade pitch: This is used to capture the optimal amount of wind while the rotor is turning.

Digital Points

Digital points are where measurements give two-position readings—such as on/off—and include the following:

- Wind direction
- Generator overheating alarm
- Hydraulic pressure level and alarm signal
- Vibration levels and alarms
- Emergency braking circuit
- Overheating of controls for yawing, hydraulic pumps, and so forth

Tech Tip
DDC Commissioning

With direct digital controls (DDCs) becoming more complex, specification engineers are requiring that commissioning be done by a third party. This means that someone other than the controls manufacturer or installation contractor must perform the final system review after the project is completely installed. This validates that the controls are fully functional and meet the specification guidelines.

PROPER SITE SELECTION

As mentioned earlier, proper site selection is inherently important to the success of the wind turbine's performance. An ideal location would be several miles offshore of one of the Great Lakes, or in the ocean. Here the winds maintain a more consistent speed and direction. However, these locations may not be the most feasible or practical from an economical or environmental standpoint. Therefore, there are a number of factors to take into consideration when choosing a site on land to erect a large-scale wind turbine. These factors are discussed next.

Roughness of the Terrain

Obviously, a smoother, more consistent terrain will create more uniform winds, whereas a rougher terrain will result in slower wind speeds (Figure 8-6). In addition, trees, shrubs, bushes, and tall grasses all contribute to the lessening of wind speeds. The wind industry categorizes the roughness of the terrain when evaluating site conditions. A roughness class of 3 to 4 refers to terrains with numerous trees and buildings. The surface of a large body of water would be considered class 0, and flat open landscapes with low amounts of vegetation receive a class score as low as 0.5.

Wind Conditions and Wind Shear

As mentioned earlier, average wind speeds for a given area can be obtained from the National Wind Technology Center website, from the U.S. Department of Energy Wind Technology Center website, or from the local meteorological center located at the nearest municipal airport. Average wind speeds are often measured by local meteorological observations at a height of about 30 feet. This should be

Figure 8-6

A smooth terrain will result in more uniform winds.

taken into consideration, due to the fact that most turbine rotors are located over 200 feet above the ground. In addition to gathering data from local sources, there are also calculators available that will assist in determining such things as wind speed, power density, and wind energy economics. **Wind shear** is another factor to take into consideration when choosing a turbine site. Wind shear is the change in wind speed with a change in altitude (Figure 8-7). As the wind's profile is twisted

Figure 8-7

Three different types of wind shear. Wind shear is the change in wind speed with a change in altitude.

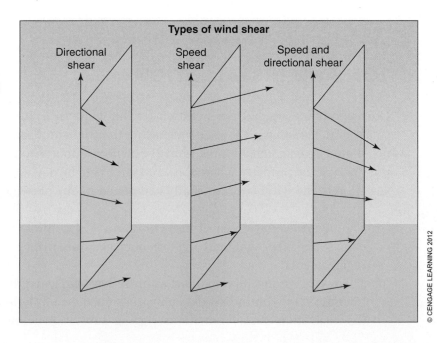

Types of wind shear

Directional shear

Speed shear

Speed and directional shear

downward toward the earth, the wind speed decreases. In fact, the wind can have a significantly higher speed at the tip of the rotor's blade when it is in its uppermost position as opposed to when it is at the bottom position of its rotation. This results in greater force acting on the blades in their top position than when the blades are in their bottom position.

Wind Obstacles

Trees, buildings, hills, and other obstacles can significantly decrease wind speeds and also create turbulence. Figure 8-8 simulates typical wind flow around an obstacle. As you can see, the turbulent area around the obstacle can extend up to three times the height of the obstacle. Also, turbulence is more pronounced behind the obstacle than in front. In addition, obstacles can decrease the wind speed downstream of them, and this decrease can be exaggerated depending upon how solid the obstruction is. For example, deciduous trees will be less solid during the winter due to the fact that their leaves are gone; however, in the summer, the tree's dense foliage will contribute to the obstruction of the wind. Therefore, it is best to try to avoid major obstacles within the area of the turbine, especially if they are upwind or in the prevailing wind direction of the turbine.

Figure 8-8

Obstacles can significantly decrease wind speeds and also create turbulence.

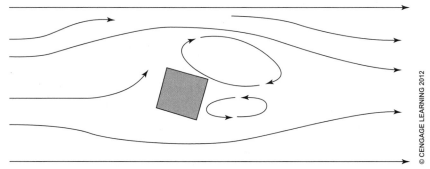

© CENGAGE LEARNING 2012

CONSTRUCTION AND INSTALLATION

Once a suitable site has been chosen for the wind farm, the construction phase can begin. This is with the understanding that the site developers have already negotiated agreements with the property owners, either through leases or through purchase of the property where the wind farm is to be developed. The actual construction of the wind farm is usually completed in one of two ways:

1. The owner or developer of the wind farm will contract with a large general contracting firm that will in turn manage the entire project and all of its requirements.
2. The owner or developer will act as his or her own general contractor and contract directly with individual companies for the completion of such things as the excavation work, foundations, rigging, substation connection, and so forth.

Before even the first shovel of dirt is moved, extensive time is devoted to the design and layout of the project. As mentioned earlier, the success of the project's design and turbine placement is determined by such things as the area terrain, prevailing winds, potential obstacles blocking wind flow, distance to the electrical substation, and any future economic development of the surrounding properties. Many key factors go into the planning of the construction phase of a wind farm, and the scheduling of the project is one of the most important areas to consider. As with any successful project, timing is of the utmost importance. One of the first tasks of the developer/general contractor is to order long-lead-time items that will be critical to the construction schedule. The actual turbines themselves along with the towers and associated equipment may have a manufacturer's lead time of weeks or even months. In addition to ordering this critical equipment in a timely manner, there are logistics that go into the actual shipping and transportation of them (Figure 8-9). For instance, the nacelles may be manufactured in Europe, the towers in Mexico, and the controls in California. It is no small feat to ship dozens of turbine blades that measure up to 150 feet in length over several countries or even across the ocean. Each of these pivotal items need to arrive on site within a specific window of time, and any delays could set the project back for days or even weeks at a time.

Once the project is mobilized, a number of tasks must be completed before even the first tower is set in place:

- **Excavation**: Roads need to be constructed to the actual tower locations. These roads need to support heavy-tonnage equipment and may stretch several miles from the main road. In addition, the tower site needs to be prepared by excavating the existing soil and by developing the foundation for the tower.
- **Trenching:** A special machine that utilizes a heavy rotating chain is used to dig trenches where underground electrical and fiber-optic cables will be buried from the tower to the substation. These trenches may stretch for many miles throughout the wind farm.
- **Substation development**: Oftentimes, a new **electrical substation** will need to be constructed at the point where the wind farm's electricity is fed into the main

COURTESY OF DONALD STEEBY

RIGUCCI, 2011. USED UNDER LICENSE FROM SHUTTERSTOCK.COM.

Figure 8-9

Transportation logistics are important to proper wind farm construction. Turbines are transported by truck or train.

Figure 8-10

Wire of this size is run underground to interconnect the wind turbines.

COURTESY OF JERRY STEEBY

grid. This area of the project will involve electrical contractors, excavation contractors, riggers, and steel erectors, as well as the local and state electric utilities, to coordinate proper completion of this phase.

Once the preliminary work is completed, other construction activities will begin. Crews will be busy completing the concrete work that is necessary for the tower foundations. Electrical contractors will lay PVC conduit through the excavated trenches and run wire from the tower locations to the electrical substation (Figure 8-10).

Another electrical crew will install pad-mounted transformers next to the towers. Meanwhile, transportation crews will deliver the primary equipment, such as the tower sections, rotor hubs, and blades, cross country while they constantly negotiate low-hanging utility wires, rough terrain, and sometimes inclement weather. Finally, the towers themselves will be set in place. Next, the nacelles and rotors will be hoisted into the air, and the final connections will be made (Figure 8-11).

Electrical Connections

Upon completion of the tower installations, there is usually a separate crew that performs the final electrical terminations and the commissioning of the new equipment. This crew may consist of authorized representatives of the turbine manufacturer or actual employees of the manufacturer. At this point, all electrical lines have been joined together and are ready for termination at the substation. In addition, fiber-optic lines have been run from each turbine's microprocessor

Figure 8-11

Once the tower is in place, the nacelle is installed and then the rotor and blades.

© ISTOCKPHOTO/MICHAEL UTECH

controller and linked together underground to a destination where a dedicated company will oversee the day-to-day operation of the wind farm. Commissioning of the control equipment is usually performed by a highly specialized group of technicians who are trained in the areas of digital controls and programmable logic. Energizing the wind farm and bringing it online is the final step in the construction process and is usually a very exciting time.

SAFETY PRACTICES FOR LARGE-SCALE WIND TURBINES

One of the most important aspects of wind turbine construction and operation is the need for safety (Figure 8-12). There are many potential hazards that face the wind turbine technician, and each of these requires the proper training and equipment in order to keep personnel safe.

Among the various safety issues that wind technicians can encounter are the following:

- Electrical shock
- Getting caught in mechanical equipment (moving parts)

Figure 8-12

One of the most important aspects of wind turbine construction and operation is the need for safety.

COURTESY OF ISTOCKPHOTO/JOERG REIMANN

- Inclement weather
- Fall hazards
- Working in confined spaces
- Strenuous work

Wind turbine workers and technicians should always adhere to local, state, and national safety standards set forth by such organizations as the Occupational Safety and Health Administration (OSHA). Furthermore, employers should require that all personnel participate in safety training courses. Regarding electrical safety, technicians should be familiar with and adhere to the proper procedures for:

- Lock-out/tag-out practices of electrical devices and disconnects
- Working around and with high-voltage equipment
- Wearing proper protective clothing and safety equipment
- Disconnect/de-energize procedures for high-voltage equipment
- Special hazards associated with high-voltage equipment
- Clearance requirements for high-voltage equipment

One of the most important requirements for technician safety is the use of proper equipment for climbing, working at heights, and working around moving equipment. Equipment for these requirements includes: proper climbing gear, hard hat, gloves, safety glasses, and steel-toed shoes. Part of proper climbing gear is ensuring that the technician has adequate fall protection. This includes a full-body harness, such as that shown in Figure 8-13, and knowledge of identifying correct anchorage points.

© ISTOCKPHOTO/JOHN TOMASELLI

Figure 8-13

Part of the technician's proper climbing gear includes a full-body harness.

Additional fall protection equipment may require energy-absorbing lanyards or fall-arrest systems. Working at heights requires practice using ladders and other climbing devices. Overall, the technician needs to know the proper procedure for both ascent and descent control.

In addition to the above-mentioned safety procedures, wind turbine technicians should have a working knowledge of the proper tower rescue procedures in the event that a worker becomes stranded or injured atop a wind turbine tower. Safety training for these procedures is available through such organizations as OSHA, the American Wind Energy Association (AWEA), and through the tower manufacturer.

Tech Tip
Worker Safety

Turbine technicians not only need to work safely, but also must be physically in good shape. Considering the fact they must climb ladders that reach over 200 feet in the air—sometimes on a daily basis—technicians must stay in top physical condition!

WIND TURBINE MAINTENANCE

Whenever there are moving parts within a machine, there is going to be wear and tear. The same is true for large-scale wind turbines. This is why an ongoing preventative maintenance program is crucial to keeping all parts in top working order, especially in cold climates (Figures 8-14 and 8-15). With wide variations in wind speed and direction, load variations, extreme changes in climate, and limited space inside the nacelle, it's easy to see why wind turbines require a sound maintenance program.

Scheduled Maintenance

Many manufacturers of large-scale wind turbines offer ongoing maintenance services and preventative maintenance contracts. These services not only result in better-working equipment, but also ensure that the equipment is up to engineering specifications. Under these services, companies generally perform comprehensive annual and semi-annual inspections of the entire turbine. In addition, other scheduled maintenance includes blade and gearbox inspection, checking yaw

COURTESY OF ISTOCKPHOTO/CAROLE CASTELLI

Figure 8-14

Ongoing preventative maintenance programs are crucial to keeping all parts of the wind turbine in top working order.

KODDA, 2011. USED UNDER LICENSE FROM SHUTTERSTOCK.COM.

Figure 8-15

Wind turbines require routine maintenance, even in winter.

and blade angles, and a review of the technician's safety program (Figure 8-16). Yaw is the right or left directional movement of the nacelle when it rotates on its vertical shaft.

Alignment Issues

A turbine that is out of alignment with the wind creates uneven loads on the blades. As a result, gusting winds can cause undue flexing and fatigue stress. This is why checking the yaw angles is so important and should be a part of the maintenance program. Some companies even offer laser alignment services for proper yaw and blade angles. It should be pointed out that for every degree that the turbine is out of alignment, the power output drops by 1%. As turbines get larger, loads are increased and the problem becomes even greater. Unless alignments of the blades and yaw are corrected in a timely manner, the average wind turbine could be out of alignment for most of its life.

Figure 8-16

Scheduled maintenance is an important part of wind turbine performance.

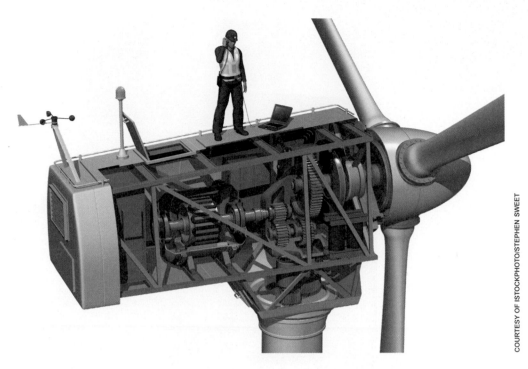

Grid Connections

Though the majority of maintenance attention is paid to the turbine itself, utility connections are also a very important part of the overall maintenance program. Typically, each turbine has its own transformer mounted at the base of the tower (Figure 8-17).

Wind turbines usually generate electricity at either 575 volts or 690 volts, which is then stepped up to 35.5 kilovolts by the transformer. These pad-mounted transformers should have their oil checked for such things as dissolved gases by taking samples on an annual basis. From the transformer, power is run to interconnecting substations, where it meets up with other turbine cabling, then on to the main substation, where it is connected to the main utility (Figure 8-18).

Cabling circuits, both electrical and fiber optic, need to have any spliced connections routinely checked to ensure that they are properly secured.

Gearbox Maintenance

It is one of the most important devices on the wind turbine, yet possibly one of the most neglected. It is the gearbox—the device that increases the RPM of the drive train located between the rotor and the generator. Estimates show that the main reason for reductions in energy output on wind farms is due to poor gearbox maintenance (Figure 8-19). On average, operation and maintenance costs begin to escalate by the fourth year of operation, with gearboxes being the biggest culprit. Although the turbines themselves have an average lifespan of 20 to 30 years, gearbox warranties may be as short as 2 years. The biggest issue with regard to gearbox problems is bearing failure. A recent study concluded that the majority of gearbox

Figure 8-17

Each wind turbine typically has its own transformer mounted at the base of the tower.

COURTESY OF DONALD STEEBY

Figure 8-18

Electric power is run to interconnecting substations, where it is connected to the main utility.

COURTESY OF DONALD STEEBY

failures begin with the bearings. This is due to a number of items, including poor lubrication and the quality of the lubricants being used. In addition to bearing issues, oil contamination can also lead to gearbox failures. This issue is caused by water and particles seeping into the gearbox crankcase, resulting in contaminated oil. The best practice for reducing oil contamination is to use proper oil filtering.

Figure 8-19

Gearbox maintenance is very important to ensuring proper performance.

Figure 8-20

An example of a 2-stage planetary gearbox.

Wind turbines utilize both inline and offline particulate filters as well as breathers or vents. Surprisingly, particulates that are invisible to the naked eye can contaminate oil and cause premature failure of gearboxes as large as 20 tons. Particle contamination can also reduce the life of gear lubricants as well. When water is allowed to find its way into the gearbox oil, it can create acids, which in turn will lead to oxidation and corrosion of the gears, bearings, and seals. The three key components to maintaining the optimum condition of gearbox oil are to keep it cool, dry, and clean. Gear oil cooling is almost as important as filtration. To help keep the oil cool, heat exchangers are used on larger gearboxes; these transfer the heat out of the gear oil prior to filtering it back into the gearbox. In order to keep the gearbox oil dry, additives are used that have a strong affinity for water, such as desiccants. These additives are mixed with the oil and the two become miscible in the crankcase. Correct filtration will ensure that the oil maintains its proper level of cleanliness. It is a good practice to install an oil-pressure switch that will read the differential pressure across the filter. This will indicate any flow restrictions across the filter due to a problem with the oil, or may indicate that it is time for the filter to be changed (Figure 8-20).

Case Study

—John Deere Wind Park, Ubly, Michigan

Travel to the "Thumb" of Michigan and you will find acres and acres of rich, flat farmland. This area is noted for its large-scale production of soybeans and sugar beets. In fact, so many sugar beets are produced in Michigan's Thumb area that the Pioneer Sugar Company has several factories located there just to handle the volume. Also originating in the Thumb area is Michigan Wind 1, the largest wind farm in the state.

Located in Ubly, Michigan, and owned by John Deere Renewables, the wind park consists of 46 GE turbines, each housing a 1.5-megawatt generator. The total capacity of Michigan Wind 1 is 69 megawatts—enough electricity to power all of Huron County. It has been in commercial operation since December 2008.

John Deere Renewables is located in Johnston, Iowa, and acquired the wind farm from Noble Environmental Power in October of 2009. Michigan Wind 1 is John Deere Renewables' second wind farm in northwest Michigan. The Harvest 1 wind farm is also in Huron County between the communities of Pigeon and Elkton. John Deere Wind Energy currently has more than 27 wind farms located in six states generating more than 600 megawatts of electrical power.

COURTESY OF DONALD STEEBY

Figure 8-21

The base of a wind turbine.

COURTESY OF DONALD STEEBY

Figure 8-22

Wind park in Michigan.

UNIT 3

Geothermal: Using the Earth to Heat and Cool Our World

HOW A GEOTHERMAL SYSTEM WORKS

Geothermal energy is one of the most sustainable sources of alternative energy and can provide a lower operating cost than any conventional type of residential or commercial heating and cooling system available. The concept of geothermal technology is to utilize the earth's crust as a medium for transferring heat. The first law of thermodynamics states that energy cannot be created or destroyed, but can be changed from one form to another. Geothermal systems utilize this principle by transferring heat to and from the earth in order to maintain a comfortable environment within a given structure. This heat transfer takes place by means of a loop of buried pipe or polyethylene tubing filled with water and antifreeze. This ground loop is referred to as the geothermal system's heat exchanger and is classified as a closed-loop system (Figure 9-1). A geothermal system may use water from an underground well or from a pond as a method of heat exchange as well.

The temperature of the earth 4 to 6 feet below its surface stays relatively stable year round. Typically, this temperature is between 50°F and 80°F, depending upon the location's annual climate. The warmth that is found in the earth is basically heat that is stored from the sun's radiation. In the wintertime, a geothermal system can extract heat from the earth and transfer it into a home or building to keep it warm. Conversely, a geothermal system can remove the excess heat from inside of a home or building and transfer that heat back into the ground

Figure 9-1

A geothermal closed-loop system for heating and cooling.

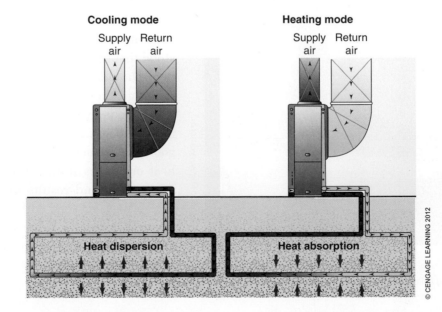

© CENGAGE LEARNING 2012

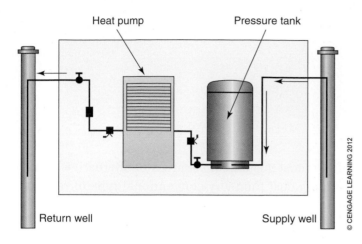

Figure 9-2

An open-loop system for geothermal heating and cooling.

Figure 9-3

Geysers, such as Old Faithful, are examples of geothermal energy found in the earth.

in the summer as a means of air conditioning the structure. Just as the earth's temperature stays relatively consistent at a depth below 4 to 6 feet, so is the temperature of water that is extracted from a well or aquifer. The type of system that uses well water is referred to as an open system. Instead of using a closed loop of buried piping, water is pumped directly from a well into the system, where the heat exchange takes place, and then is pumped back into the ground (Figure 9-2).

As mentioned earlier, geothermal systems have the capability of heating structures during the winter because of the sufficient amount of heat found in soil or water that has a temperature as low as 50°F. This is because the water that is used in the closed-loop heat exchanger or from a well has the capability to store and transfer a tremendous quantity of heat. These two facts combine to make geothermal heat pumps one of the most efficient sources of heating and air conditioning available. The efficiency of a geothermal system lies in the fact that it transfers heat rather than produces heat, such as by the burning of fossil fuels. This heat transfer is done by means of the system's self-contained refrigeration system.

Geothermal heat pumps are known by a variety of other names, including **geoexchange**, earth-coupled heat pumps, and water-source heat pumps. It should be understood that the type of system being discussed here should not be confused with conventional geothermal energy. Conventional geothermal energy traditionally refers to the type of heat that is found deep within the earth's geological structure. This type of geothermal heat originates from sources such as volcanic activity and results in the formation of hot springs, geysers, and naturally occurring steam (Figure 9-3). These sources are used for applications such as powering large turbines to produce electricity.

Things to Know

GEOTHERMAL ENERGY VERSUS GEOTHERMAL HEAT PUMPS

Conventional geothermal energy originates from the radioactive decay of minerals, and from volcanic sources deep within the earth's surface. This type of energy is expressively displayed by examples such as the geyser "Old Faithful" in Yellowstone Park.

The geothermal energy that is used for residential and commercial heating and cooling is a result of transferring heat into and out of the earth's crust. This task is accomplished by the use of a ground-source heat pump (GSHP).

THE GEOTHERMAL REFRIGERATION SYSTEM

A typical room air conditioner or household refrigerator consists of a basic refrigeration system. The purpose of air conditioning is to remove unwanted heat from the home or building by absorbing this heat into the refrigerant-filled indoor coil and transferring it to the outdoor coil, where it is removed. A conventional air-source heat pump uses this same system for air conditioning but reverses the refrigeration system when it is being utilized in the heating mode. When operating in the heating mode, an air-source heat pump is absorbing heat from the outdoor air into the refrigerant-filled outdoor coil and transferring this heat into the home or building. A geothermal heat pump essentially works in this same manner except there is no conventional outdoor coil. Instead, heat is transferred by pumping water and antifreeze through a ground loop series of piping or by means of water that is pumped out of a well. This water or antifreeze is circulated through a heat exchanger mounted inside the heat pump, where the heat transfer takes place.

As shown in Figure 9-4, the basic refrigeration system consists of four essential devices: the compressor, the condenser, the metering device, and the evaporator. No matter what size the refrigeration equipment, every system is required to include these four elements in order to work, and each has its own distinct purpose.

The Compressor

The **compressor** is the heart of the refrigeration system. Its main purpose is to circulate refrigerant through the system. However, it essentially performs two functions. First, it draws refrigerant vapor out of the evaporator, and second, it compresses the refrigerant. This in turn raises its temperature and pressure. By raising the temperature of the refrigerant, heat transfer can occur due to the fact

Figure 9-4

The four essential objects required for the refrigeration cycle.

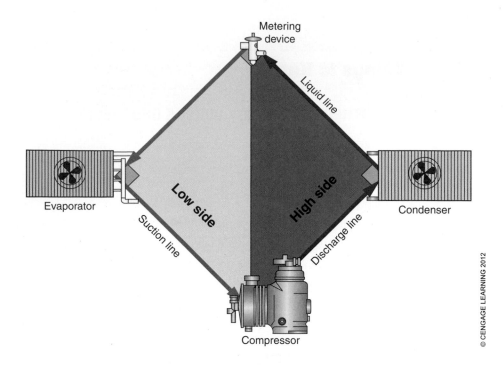

Metering device

Liquid line

Evaporator

Condenser

Low side

High side

Suction line

Discharge line

Compressor

© CENGAGE LEARNING 2012

that heat naturally flows from warm to cold. Most ground-source heat pumps are equipped with either reciprocating or scroll-type compressors.

Reciprocating compressors work similarly to an internal combustion engine (Figure 9-5). They consist of pistons and valves that draw refrigerant into the cylinders and in turn compress it to raise the temperature and pressure (Figure 9-6).

© CENGAGE LEARNING 2012

Figure 9-5

The internal workings of a reciprocating compressor.

COURTESY OF DONALD STEEBY

Figure 9-6

A cut-away view of a reciprocating compressor.

The other main type of compressor used with geothermal heat pumps is known as a scroll-type compressor (Figures 9-7 and 9-8). This type consists of stationary and orbital scrolls that compress the refrigerant as the scrolls rotate around each other (Figure 9-9). Figure 9-10 shows a cut-away view of a scroll compressor. Notice that the interlocking scrolls are positioned in the dome of the compressor.

Figure 9-7

A modern scroll-type compressor.

COURTESY OF DONALD STEEBY

© CENGAGE LEARNING 2012

Figure 9-8

Internal view of a scroll-type compressor.

Figure 9-9

How the refrigerant is compressed through a scroll compressor.

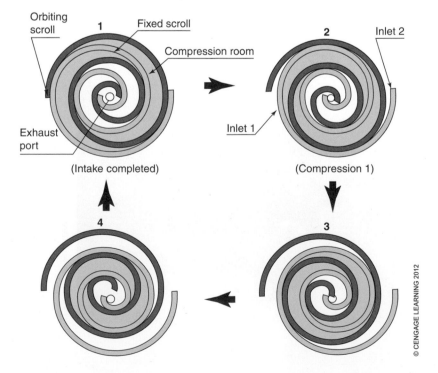

© CENGAGE LEARNING 2012

Figure 9-10

A cut-way view of a scroll compressor showing the interlocking scrolls located in the dome.

COURTESY OF DONALD STEEBY

Almost all compressors used in conjunction with geothermal heat pumps are hermetically sealed and are not field serviceable. They rely on the cooling effect of the refrigerant vapor to keep the bearings and other components cool under extreme conditions (Figure 9-11).

Figure 9-11

A welded hermetic compressor.

COURTESY OF DONALD STEEBY

The Condenser

Once the refrigerant leaves the compressor it is pumped into the **condenser**. This is where the superheated refrigerant vapor gives up its heat and condenses into a liquid. By definition, superheat is the temperature of the refrigerant above the point when it has become 100% saturated vapor. Most conventional air conditioning condensers are air-cooled (Figure 9-12). However, in a geothermal heat pump,

Figure 9-12

An air-cooled condensing unit.

COURTESY OF DONALD STEEBY

the condenser is actually a coaxial coil consisting of a tube inside of another tube. One tube contains the refrigerant and the other contains the water and antifreeze solution that is circulated through the ground loop (Figure 9-13). These two substances are pumped in opposite directions through the coaxial heat exchanger to

Figure 9-13

A coaxial-type heat exchanger. This device acts as the condenser in the cooling mode on a geothermal heat pump.

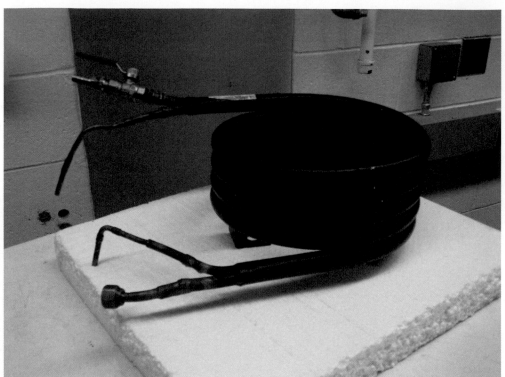

COURTESY OF DONALD STEEBY

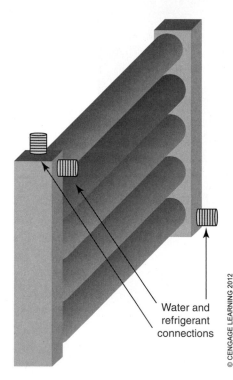

Water and refrigerant connections

© CENGAGE LEARNING 2012

Figure 9-14

A tube-within-a-tube heat exchanger constructed by sliding one tube through another. The tubes are sealed in such a manner that the inside tube is separated from the outside tube.

maximize heat transfer. This process is known as counterflow heat exchange (Figure 9-14).

The condenser serves three main purposes. First it de-superheats the hot vapor that is pumped from the compressor. Most of this heat is generated as a result of compressing the refrigerant; however, some is a result of the heat generated from friction within the compressor—much like the heat buildup inside an internal combustion engine. The second purpose of the condenser is to actually condense the refrigerant vapor into a liquid. This process is essential to heat transfer. As the refrigerant condenses, it changes state, and therefore removes the latent heat from the system. By definition, **latent heat** is the heat energy that is absorbed or rejected when there is a change from one state to another, in this case from a vapor to a liquid. The third purpose of the condenser is to subcool the liquid refrigerant. Subcooling is the additional sensible heat that is removed from the refrigerant once it has become 100% saturated liquid. Subcooling is an important aspect of the refrigeration system because it is used to determine the proper refrigerant charge of the system and also determines the effectiveness and capacity of the metering device.

The Metering Device

The purpose of the metering device is to dramatically reduce the pressure of the refrigerant, which in turn reduces its temperature. A full column of liquid refrigerant is delivered to the metering device from the condenser by means of the liquid line, which runs from the outlet on the condenser. There are mainly two types of metering devices. One is classified as a fixed-orifice device and the other is known as an **expansion valve**. The **capillary tube** is the typical fixed-orifice metering device found on geothermal heat pumps. As the name implies, the capillary tube is a long, small copper tube that reduces the pressure of the liquid refrigerant as it is passed through (Figure 9-15). The capillary tube is connected between the liquid line and the evaporator. The **thermostatic expansion valve**, or **TXV**, is the other common type of metering device. In fact, TXVs are found almost exclusively on all modern-day geothermal heat pumps (Figure 9-16).

Figure 9-15

A capillary-tube-type metering device connected between the liquid line and the evaporator.

COURTESY OF DONALD STEEBY

Whereas the capillary tube delivers a constant volume of refrigerant to the evaporator, the TXV can modulate the flow of refrigerant, making it a more efficient device. This modulation is determined by the load on the evaporator. For instance, during air conditioning, if the temperature of the conditioned space is well above the desired setpoint, the TXV will throttle open to deliver a larger volume of refrigerant to the evaporator in an attempt to overcome the large cooling load on the space. As the space begins to cool down and approach the desired space temperature setpoint, the TXV begins to close so as not to over-feed refrigerant to the evaporator, causing it to flood (Figure 9-17).

Figure 9-16

A thermostatic expansion valve found on the indoor coil of a heat pump. This device meters the refrigerant into the coil.

COURTESY OF DONALD STEEBY

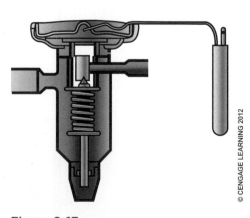

Figure 9-17

A cut-away view of a thermostatic expansion valve. Notice the needle valve that modulates to control the flow of refrigerant into the evaporator.

© CENGAGE LEARNING 2012

Whether the capillary tube or TXV type of metering device is incorporated into the heat pump, the liquid refrigerant passes through a small restriction within the metering device that causes it to change from a high-temperature, high-pressure liquid into a low-temperature, low-pressure liquid-vapor mix. This mixture resembles a dense fog as it is sprayed into the evaporator. It is important to remember that the metering device performs double duty on the geothermal heat pump system. This is because the heat pump is used for both heating and cooling. In the heating mode, the metering device modulates the flow of refrigerant to the coaxial heat exchanger. This is because the coaxial heat exchanger acts as the evaporator during the heating mode. In the cooling mode, the metering device modulates the flow of refrigerant to the coil located in the furnace plenum. This indoor coil acts as the evaporator in the cooling mode. In some cases, check valves may be installed along with the metering devices to ensure that refrigerant can flow in only one direction. Figure 9-18 shows the direction of refrigerant flow through the heat pump during the heating mode. Figure 9-19 shows the direction of flow during the cooling mode.

Figure 9-18

The direction of refrigerant flow with the heat pump in the heating mode.

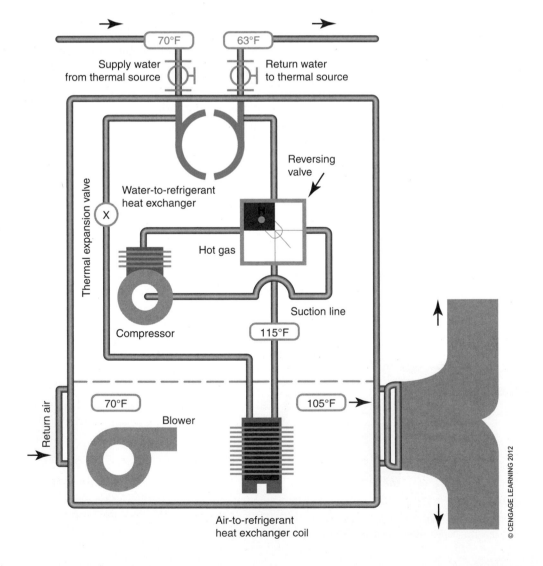

© CENGAGE LEARNING 2012

Figure 9-19

The direction of refrigerant flow with the heat pump in the cooling mode.

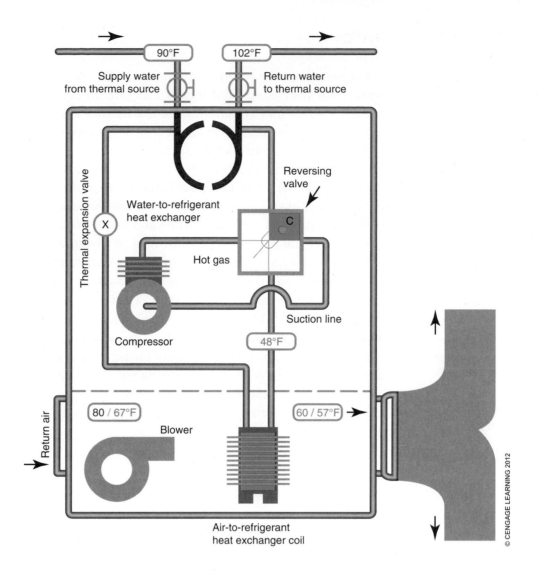

Tech Tip

Adjusting the Thermostatic Expansion Valve

When trying to squeeze more heating or cooling out of the geothermal heat pump, there may be a tendency to want to change the setting on the superheat adjustment screw located under the thermostatic expansion valve. It would be wise to resist this temptation. Many service technicians have caused more harm than good by attempting to change this setting. If there is an issue with the superheat measurement at the evaporator coil, it can usually be remedied by something other than changing the factor adjustment of the superheat spring.

The Evaporator

The **evaporator** is the fourth component in the basic refrigeration system. On a geothermal heat pump, the indoor coil is considered the evaporator during the cooling mode and is found inside the cabinet along with the system circulating fan (Figure 9-20). The evaporator acts the same as a sponge by absorbing heat from the conditioned space during the cooling mode, and from the ground during the heating mode. When the refrigerant enters the evaporator, it consists of a low-temperature, low-pressure, liquid-vapor mix. Warm, humid air is circulated across the evaporator coil from the conditioned space. As it passes through the coil, this air gives up both sensible and latent heat, which is absorbed by the refrigerant. As the refrigerant absorbs heat, it changes from a liquid-vapor mix to all vapor. An enormous amount of heat is transferred during this refrigerant phase change of liquid to vapor, and this process is known as the latent heat of vaporization. Once the refrigerant state is 100% vapor, it has the capability to absorb sensible heat. The amount of sensible heat that increases the temperature of the refrigerant is known as superheat. Superheat plays an important role in the refrigerant system. It is an accurate measurement in determining the proper charge in a fixed-orifice system. It also is used to ensure that no liquid refrigerant is allowed to migrate back to the compressor. The compressor is designed to pump vapor only, not liquid. If any liquid refrigerant finds its way back to the compressor, irreparable damage may occur.

Figure 9-20

The evaporator is contained in the indoor cabinet of the geothermal heat pump.

The Four-Way Reversing Valve

Figure 9-21

A four-way reversing valve.

There is a fifth element that is not required for the basic refrigeration system to operate properly; however, it is specific to heat pump applications: the four-way **reversing valve**. This valve is utilized in geothermal heat pumps to reverse the flow of refrigerant between heating and cooling modes (Figure 9-21).

The four-way reversing valve consists of four ports of tubing that connect to the different components of the heat pump. Three of the ports are on one side of the valve, and the fourth is on the opposite side. The middle of the three ports is permanently connected to the suction or inlet line of the compressor. The single port is permanently connected to the discharge line of the compressor. Of the two remaining ports, one is connected to the line leading to the coaxial coil and one is connected to the line leading to the indoor coil. The four-way valve also has a solenoid pilot valve. This solenoid valve changes the direction of refrigerant flow when it is energized, either in the heating or cooling mode, by moving an internal slide mechanism (Figure 9-22).

Most heat pump manufacturers install the four-way valve so that it is in the heating mode when the solenoid is de-energized. There are also small pilot lines that run from the single port to the solenoid pilot valve and then to each end of the main valve body. These pilot lines use the pressure of the refrigerant to assist in moving the slide mechanism when the valve switches from one mode to the other.

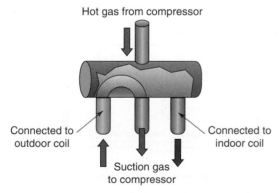

Four-way valve in the heating mode

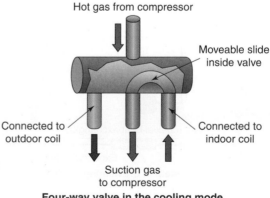

Four-way valve in the cooling mode

Figure 9-22

Direction of refrigerant flow through the four-way reversing valve in the heating mode and cooling mode.

Field Tip

Replacing the Four-Way Reversing Valve

Although reliable for the most part, sometimes the four-way reversing valve needs to be replaced. One of the issues with **brazing** a new valve in place is the fact that the three ports are very close together, making brazing a challenge in the cramped quarters inside the heat pump. One solution in making this task easier is to first braze copper tubing extension onto the individual ports of the new valve before installing it in the heat pump cabinet. This will allow for more room when brazing the final joints, and help to assure that they have been brazed successfully.

Remember that the flow of refrigerant is not reversed through the compressor. A compressor pumps vapor in only one direction. However, the refrigerant that is discharged from the compressor can be directed to either the indoor coil or the outdoor coaxial coil. This is where the four-way reversing valve is utilized. In the heating mode, the heat pump's refrigerant flows from the compressor directly to the indoor coil. This coil acts as the condenser in the heating mode, where it dispenses heat into the conditioned space when there is a call for heating. Conversely, the outdoor coaxial coil is acting as the evaporator when the heat pump is in the heating mode. The ground loop is absorbing heat from the earth, where it will be transferred into the refrigerant located inside the coaxial heat exchanger. This heat will warm the conditioned space during the winter.

In the cooling mode, the refrigerant flows from the compressor directly to the outdoor coaxial coil. This coil acts like the condenser in the cooling mode, transferring the heat into the ground loop. Here the ground loop rejects this heat that originated in the conditioned space into the earth. In the heat pump cooling mode, the indoor coil is the evaporator, removing heat from the conditioned space. It is important to review Figures 9-18 and 9-19 for a clear understanding of refrigerant flow through the heat pump in both the heating and cooling modes.

Figure 9-23 show another view of the system components within the geothermal heat pump. Notice the locations of the coaxial heat exchanger and the indoor coil. Remember that the reversing valve, expansion valve, and coaxial heat exchanger are all packaged within the heat pump cabinet.

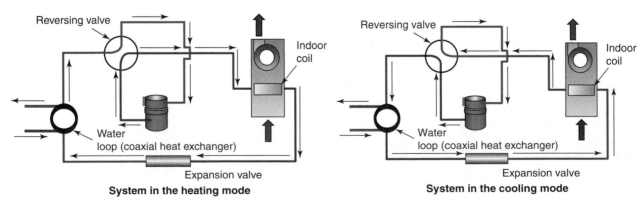

© CENGAGE LEARNING 2012

Figure 9-23

An overall view of the direction of refrigerant flow through the geothermal heat pump in the heating mode and cooling mode.

Chapter 10

TYPES AND CONFIGURATIONS OF GEOTHERMAL LOOPS

Geothermal loops can be designed with a variety of configurations. The particular configuration that is chosen will depend on a number of factors for the specific application. As stated previously, the two main types of ground loops are classified as open loops and closed loops.

An open-loop configuration is where water is pumped from an underground aquifer, pond, lake, or stream through the coaxial heat exchanger and discharged back into the ground. When available, ground water from an aquifer is the preferred choice for an open-loop system (Figure 10-1).

When choosing an open-loop system, it is best to use a dedicated well for the geothermal heat pump, rather than utilize an existing domestic well, because of the volume of water used by the heat pump. One advantage in choosing an open-loop system is that the temperature of the water in the aquifer is generally consistent. The three main factors to consider when choosing this type of configuration are: water quality, water quantity, and where to pump the water once it has been circulated through the heat pump—in other words, where to get rid of it.

Figure 10-1

Aquifers are geological formations formed from underground wells or springs.

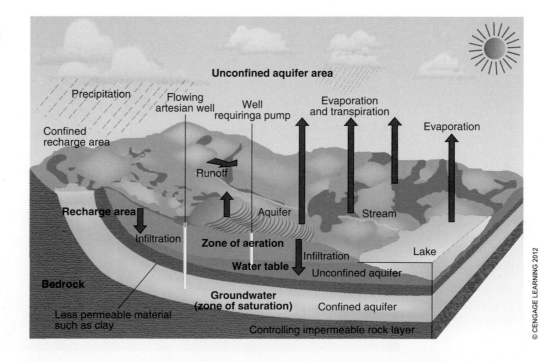

OPEN-LOOP WATER QUALITY CONSIDERATIONS

Ground water is produced from underground wells or springs and is found in geologic formations known as aquifers. The water from these sources is utilized in a geothermal open-loop system and continuously recycles itself through a process called the hydrologic cycle (Figure 10-2).

The three stages of the hydrologic cycle consist of evaporation, condensation, and precipitation. Water is first evaporated by the heating of the sun. Then it vaporizes and becomes part of the earth's atmosphere. As the water vapor condenses, it forms clouds. When the moisture in the clouds increases to a point where the clouds can no longer hold it, they give up precipitation in the form of rain or snow. A portion of this water is infiltrated back into the earth's aquifers, where it is utilized by the open-loop system. As water passes through the earth, it dissolves minerals such as gypsum, limestone, calcium carbonate, and other forms of sedimentary rock. These minerals are absorbed into the water, resulting in poor water quality such as hardness and incrustation from iron. If left unchecked, this situation can lead to scaling and corrosion within the heat pump's heat exchanger (the coaxial coil mentioned in Chapter 9) and reduce the thermal conductivity of heat transfer (Figure 10-3).

Figure 10-2

The hydrologic cycle provides the ground water used in open-loop geothermal systems.

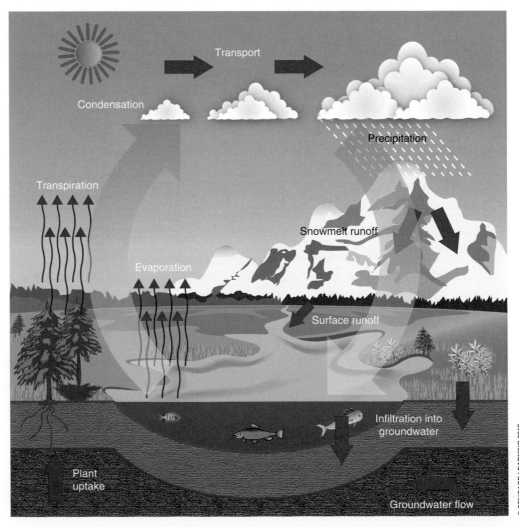

© CENGAGE LEARNING 2012

Figure 10-3

Heat exchanger buildup.

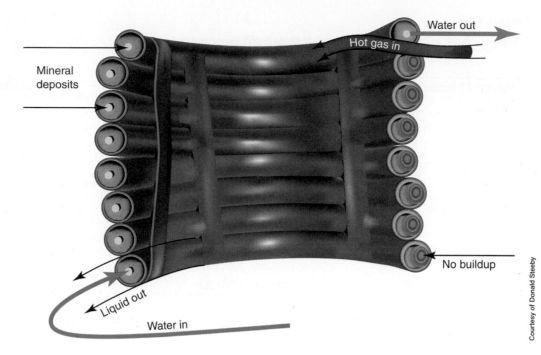

Mineral deposits

Water out

Hot gas in

No buildup

Liquid out

Water in

Courtesy of Donald Steeby

All of these factors can be detrimental to the lifespan of the open-loop heat exchanger. For these reasons, it is important that the well water be tested for its quality before the decision is made to pursue an open-loop type of system.

The heat exchanger in an open-loop geothermal heat pump is made of either copper or cupronickel. Copper heat exchangers are more acceptable for closed-loop systems, where the quality of the water can be controlled. Coaxial heat exchangers that are made of cupronickel will have a higher resistance to abrasion and corrosion than those made of straight copper. They are more conducive to open-loop systems and also can withstand acid cleanings. Table 10-1 can be used as a guideline for determining which type of material is best used with an open loop heat pump system.

Things to Know

WATER HARDNESS RATINGS

Hard water is the result of dissolved minerals such as calcium and magnesium. Hard water requires more detergent in order to get clothes clean, and also contributes to scaling in the heat exchanger. The hardness factor in water is rated in grains per gallon. Water that contains less than 1 grain per gallon of calcium or magnesium is considered "soft," whereas water that contains more than 10 grains per gallon is classified as "very hard."

Table 10-1

Selection Guide to Compare Heat Exchangers Made from Copper and Cupronickle

WATER COIL SELECTION GUIDE		
Potential Problem	**Use Copper Coil**	**Use Cupronickle Coil**
Scaling		
Calcium and magnesium salts	If less than 350 ppm	If more than 350 ppm
Iron oxide	If low concentration	If high concentration
Corrosion*		
pH Levels	If pH is 7 to 9	If pH is 5 to 7 or 8 to 10
Hydrogen sulfide	Less than 10 ppm	10 to 50 ppm
Carbon dioxide	Less than 50 ppm	50 to 75 ppm
Dissolved oxygen	Only with pressurized water tank	With all systems
Chloride	Less than 300 ppm	300 to 600 ppm
Total dissolved solids	Less than 1,000 ppm	1,000 to 1,500 ppm
Biological Growth		
Iron bacteria	Low	High
Suspended Solids	Low	High

*Notice – If the concentration of these corrosives exceeds the maximum levels in the cupronickel column, water treatment may be required.

Water softeners may be used to prevent water hardness, and the use of filters can reduce the amount of iron buildup. Unfortunately, the use of resin-bed water softeners is not recommended with geothermal heat pumps due to the fact that it is not economically practical and can also deteriorate the heat exchanger. This is why it is important to have the water source tested for hardness and for iron content. If the water quality does not meet the standards of the heat pump manufacturer, then an alternative closed-loop design may be required. If the quality of the water is marginal, there are methods to reduce the amount of scaling within the heat pump. These include reverse-cycling the unit or by chemical cleaning. Reverse-cycling changes the heat exchanger to a low-temperature evaporator. Performing this procedure results in lowering the temperature of the water within the heat exchanger to allow more carbon dioxide into the water, thus removing scale buildup. Chemical cleaning of the heat exchanger is the preferred method of scale removal and involves back-flushing the unit with cleaning acid. A small pump circulates the solution in the opposite direction from normal flow for at least 3 hours. Once this procedure is performed, the unit is flushed with clean water for at least 10 minutes.

DETERMINING PROPER WATER QUANTITY IN THE OPEN-LOOP WELL

The rate at which water enters the aquifer depends on the local climatic conditions. The level at which the ground water is found at the top of the aquifer or its saturated zone is known as the water table. The level of the water table will vary depending on the type and permeability of the rock or soil. The water table will also rise and fall depending on the season of the year and on the amount of rain or snow melt that occurs. The amount of water that a well can produce is referred to as the yield, and depends on the size of the aquifer, the type of well that is installed, and the pumping characteristics of the well. A short-term yield depends on the capacity of the pump, and a long-term yield depends on the size and type of aquifer.

Testing the well for its capacity and pumping characteristics is an important aspect of the geothermal system and should be given close attention. Running a well dry, not to mention the neighbor's well, is not something that an owner wants to have happen.

The following characteristics are used to determine the performance of the well:

- **Static water level**: The distance from the surface of the ground to the water in the well bore or hole when there is no pumping taking place (Figure 10-4).
- **Pumping level**: The distance from the surface of the ground to the water level when the well pump is in operation.
- **Drawdown**: The distance between the static water level and the pumping level. It is the distance that the water level is reduced as a result of pumping the water (Figure 10-5).
- **Residual drawdown**: The distance between the static water level before and after the well has been pumped.
- **Well yield**: The pumping rate measured in gallons per minute.
- **Specific capacity**: Gallons per minute per foot of drawdown. This figure is calculated by dividing the yield by the drawdown distance. This characteristic should be measured after the pump has been operational for a length of time, usually 24 hours.

Figure 10-4

The static water level is the distance from the surface of the ground to the water in the well bore or hole when there is no pumping taking place.

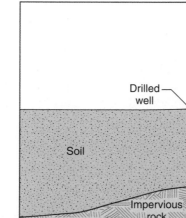

© CENGAGE LEARNING 2012

Figure 10-5

Drawdown is the distance between the static water level and the pumping level. It is the distance that the water level is reduced as a result of pumping the water.

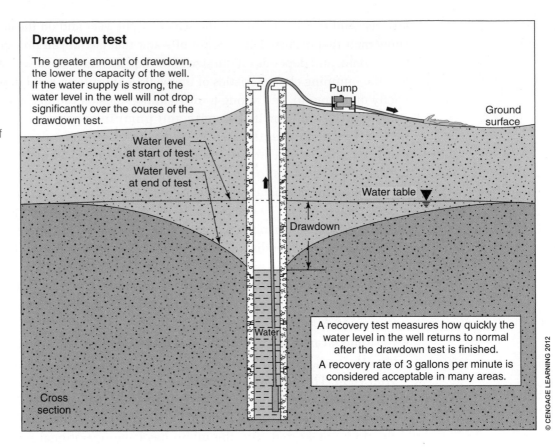

Drawdown test

The greater amount of drawdown, the lower the capacity of the well. If the water supply is strong, the water level in the well will not drop significantly over the course of the drawdown test.

Water level at start of test

Water level at end of test

Pump

Ground surface

Water table ▽

Drawdown

Water

A recovery test measures how quickly the water level in the well returns to normal after the drawdown test is finished.

A recovery rate of 3 gallons per minute is considered acceptable in many areas.

Cross section

© CENGAGE LEARNING 2012

Things to Know

CAPACITY AND TONNAGE

When describing the capacity of a heat pump in tons, remember that 1 ton of cooling is equal to 12,000 BTU/hr.

There are other capacity tests that can be performed by the well driller to help determine the capacity and pumping rate of the well. In order to achieve the proper performance of the well, the driller must know the expected requirements for the heat pump system before beginning. This includes the required gallons per minute of flow through the heat pump. On average, most heat pumps require a flow of between 3 and 5 gallons per minute per ton of capacity for the cooling mode. This may equate to a required well capacity that can produce from 6 to 25 gallons per minute of water flow. To put these numbers into perspective, a heat pump that uses 10 gallons per minute, and is operating continuously for one day, will require 14,400 gallons of water!

OPEN-LOOP CONFIGURATIONS

Once it has been determined that conditions are adequate to develop an open-loop geothermal system, several types of well configurations can be utilized. As mentioned earlier, water sources for an open-loop system may be from an existing well or from a dedicated one. Regardless of the type of well that is utilized, most will draw water from the aquifer by means of a submersible pump. This pump is usually located at the bottom of the well casing and is submerged in the aquifer's water table. The well casing is made of steel or thermoplastic piping and is used to prevent the soil from collapsing around the well, and also to prevent contamination from surface water. Also most wells are grouted. Grouting is a process in which a cement-like material such as bentonite is injected between the well casing and the bore hole. When the grout hardens, it forms a protective seal that prevents the well from becoming contaminated by any other water sources. Grouting makes the entire well structure sturdier and can prevent the well casing from rusting.

There are several different well configurations to choose from when installing an open-loop system (Figure 10-6). The most popular configurations are:

- Conventional drilled well
- Utilizing a return well
- Standing column
- Utilizing a dry well system

© ISTOCKPHOTO/CARROTEATER

Figure 10-6

The driller must know the expected requirements for the heat pump system before beginning. This includes the required gallons per minute of flow through the heat pump.

Conventional Drilled Well

Most conventional wells are dug using several different methods. These include:

- **Driven wells**: This process uses a special pointed pipe called a drive point that is pounded into the ground. The drive point has a special screen attached to it to filter out sediment. Driven wells are usually no more than 50 feet deep and produce small to moderate yields of water. They are not recommended where there are large volumes of dense rock below the surface.
- **Bored wells**: Here an auger is used to bore into the earth and carry the refuse to the surface. The casing is then inserted into the bored hole. Bored wells can be large in diameter and are used where large volumes of water are required. Like driven wells, bored types are difficult to construct in dense rock.
- **Drilled wells**: This type of drilling process is used in a variety of conditions and is performed by the use of a cable tool or rotary process. Known as the percussion process, cable tooling utilizes a heavy drill bit that is raised and dropped into the hole, loosening rock and other material (Figure 10-7).
- **Rotary drilling**: This method incorporates a rotating drill bit that breaks up the earth and mixes it with a drilling fluid to wash the mixture from the drilled hole. Once the drilling is complete, the well casing is inserted into the hole. Rotary drilling is the most recommended technique for geothermal applications.

Figure 10-7

A basic drilled well.

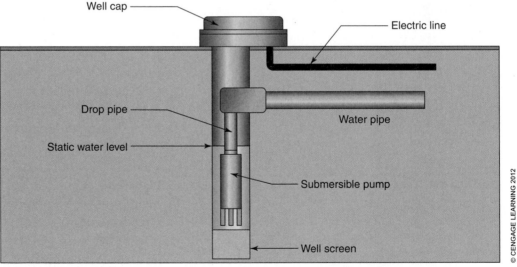

Well cap

Electric line

Drop pipe

Water pipe

Static water level

Submersible pump

Well screen

© CENGAGE LEARNING 2012

Utilizing a Return Well

A return well is used to discharge the water back into the ground once it has passed through the heat pump's coaxial heat exchanger. Figure 10-8 shows a typical return well configuration. One consideration for utilizing a return well is to be sure that the separate wells are spaced at a distance far enough to prevent mixing of the supply and return water. If they are not, the two waters could mix, causing the supply water to be too hot or cold, depending upon the season. This effect will severely reduce the capacity of the heat pump. It is recommended that the wells

Figure 10-8

A return well system.

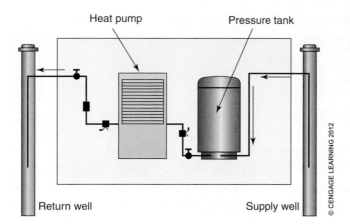

Heat pump　　Pressure tank

© CENGAGE LEARNING 2012

Return well　　Supply well

maintain a minimum distance of at least 100 feet between them. Furthermore, the return well must be at least the same size as the supply well or larger to handle the water flow.

Standing Column

A **standing column** or dedicated geothermal well is used when there is not enough water in an underground aquifer to utilize a standard well system (Figure 10-9). This type of configuration consists of the supply and return lines being piped into the same bore hole. The supply water is drawn from the top of the water column and into the heat pump. The return line is terminated at a much lower point than the supply line toward the bottom of the well. The theory is that once the return water reaches the supply water level, it will have been tempered to prevent premature mixing. A bleed valve is utilized with the standing column system that, when energized, shifts the operation

Figure 10-9

A standing column well system.

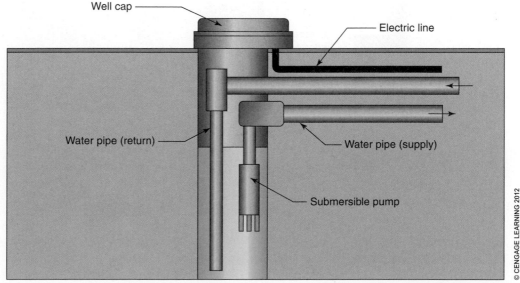

Well cap

Electric line

Water pipe (return)

Water pipe (supply)

Submersible pump

© CENGAGE LEARNING 2012

into a conventional open-loop mode when the heat pump is under peak loads. Standing column systems are typically sized at 40 to 50 feet of bore per ton of heat pump capacity.

Dry Well Systems

A **dry well** is used as a means for discharging the water in an open-loop system (Figure 10-10). Typically, dry wells are simply large reservoirs in the ground that are filled with gravel and sand. They are mostly used in areas with sandy soils. Once the return water enters the dry well, it is filtered as it seeps through the sand and gravel until it eventually returns back to the underground aquifer.

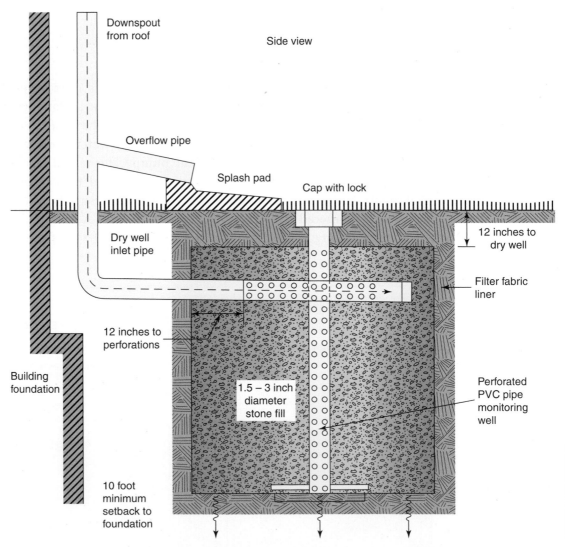

Figure 10-10

A dry well.

PRESSURE TANKS

A pressure tank is simply a small pressurized vessel used for water storage. With all open-loop-type systems, pressure tanks are utilized to prevent the pump from running every time a small amount of water is used by the system. Short cycling of the pump can result in its premature failure. Even when the pump is de-energized, the pressure tank can supply enough water to the system when there is a call for heating or cooling. The tank contains a rubber bladder inside of it that separates the air from the water. The air side of the tank contains compressed air that can be regulated by means of a fitting with which to increase or decrease the pressure, similar to filling an automotive tire with air. The system pump is activated by means of a pressure switch. Common switch settings would be to energize the pump when the system pressure falls to 20 psi, and then de-energize the pump when the pressure reaches 40 psi. Figure 10-11 illustrates the operation of the systems pressure tank.

Field Tip

Well Water Disposal

It is important to remember that if the return water from a geothermal well is discharged onto the surface of the ground, it should be done near a stream, pond, or lake. Unhappy relationships with neighbors whose backyards have flooded are a result of improper water disposal practices. Remember that the average open-loop heat pump discharges large volumes of water on a daily basis during the heating or cooling seasons. It is imperative that proper return water practices be implemented when planning this type of system.

Figure 10-11

The operation of a well system pressure tank.

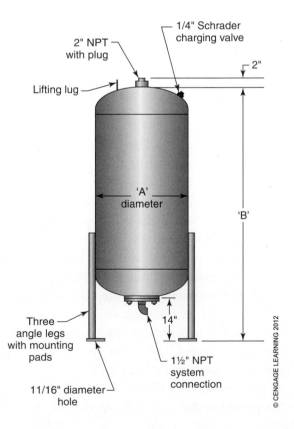

© CENGAGE LEARNING 2012

CLOSED-LOOP CONFIGURATIONS

The other main type of geothermal ground loop configuration is classified as a closed loop. The type of piping material used for most closed-loop systems is typically ¾-inch-diameter high-density polyethylene or polybutylene. This type of piping is used primarily because it is resistant to corrosion while maintaining a high level of heat-transfer capability. Also, when joints are made with this type of piping using a heat-fusion process, the joints are stronger than the original material. The piping is filled with water or a water and antifreeze solution, and completely sealed. The solution is circulated through the loop by a low-wattage centrifugal pump. Heat is transferred from the ground into the water and anti-freeze solution inside the piping. This piping is connected to the coaxial heat exchanger located in the cabinet of the heat pump, where it exchanges energy with the refrigerant loop. Remember that the coaxial heat exchanger is a tube-in-tube heat exchanger where the refrigerant flows through the outside tubing and the water loop through the inside. The two fluids flow in opposite directions to create what is called counterflow heat transfer. When compared to an open-loop type of configuration, closed loops have several advantages. One advantage is that there is no issue with regard to water disposal as compared to an open-loop system. Another is that the quality of the ground water does not affect the loop's system performance. When the quantity of ground water is an issue, a closed-loop system is the answer. There are several variations of the closed-loop system. Typical closed-loop configurations include:

- Vertical loops
- Horizontal loops
- Slinky loops
- Pond loops

Vertical Loops

When the amount of available land is limited, **vertical ground loops** are typically used (Figure 10-12). Usually a well driller is contracted to bore several holes or wells that measure up to 6 inches in diameter and have an average depth of 150 to 250 feet. The drilling procedure is similar to that of an open-loop configuration. The driller needs to be prepared for the soil conditions that may be encountered and for the type of material that is extracted from the ground, called overburden (Figure 10-13). This overburden may be anything from sand to gravel and even bed-rock, depending upon the geological characteristics of the region. Modifications may need to be made to the system depending upon the type of overburden. For instance, if there is extreme overburden encountered during the drilling process, the driller may opt for drilling four 100-foot well depths instead of just one 400-foot well. For these reasons, test holes are sometimes drilled and the soil conditions analyzed on large commercial projects.

The manager of the project also should be mindful that hydrostatic pressures increase inside the piping as the well is made deeper. At a depth of 340 feet, the

Figure 10-12

A vertical ground loop configuration.

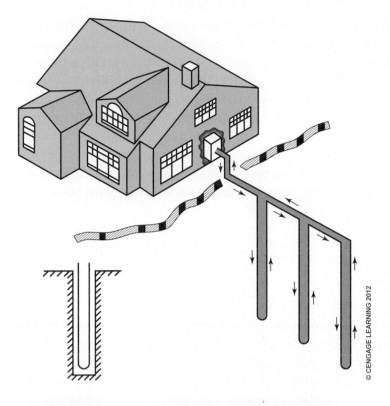

© CENGAGE LEARNING 2012

Figure 10-13

Drilling vertical loops requires heavy-duty equipment to encounter different types of overburden.

© ISTOCKPHOTO/JOHN COOKE

pressure inside the buried pipe can reach 120 psi. When combined with the pressure encountered from the circulating pump, this hydrostatic pressure can be as high as 190 psi. Keep in mind that most of the polyethylene piping used for vertical closed-loop systems is rated at 180 psi. Fortunately, this extreme pressure may not be an issue because of the external pressure that is encountered from outside of the buried pipe, creating an offsetting pressure differential.

Most common vertical configurations utilize the following parameters: ¾-inch piping, installing one loop per ton of system capacity, with each loop depth being 150 feet. Piping diameter is usually dictated by the depth of each well. The following parameters outline the required diameter of the piping based upon the depth of the well:

100–200 ft. = ¾″ pipe
200–350 ft. = 1″ pipe
300–550 ft. = 1¼″ pipe

Once the well holes are drilled, the polyethylene piping is inserted into the holes with a hairpin fitting that is connected to the end of the loop. The well hole is then back-filled with a form of grout called bentonite. This special grout material serves two purposes. It enhances the thermal conductivity of the piping and ensures that there will be no ground contamination should antifreeze ever leak out of the piping. The individual vertical loops are then connected to a common header that is also made of polyethylene piping. This header is buried in a trench that leads back to the heat pump, which is located inside the home or building.

Things to Know

VERTICAL BORE INSPECTIONS

In most states, individuals constructing wells or bore holes are required to be licensed. In addition to this, more and more states are requiring that bore holes, as well as all fused joints in vertical closed-loop systems, be inspected. This practice only makes sense. By doing so, the consumer and the installer are protected from liability as a result of leaking ground loops.

Horizontal Loops

When an adequate amount of land is available that does not contain a high level of rock and underground debris, a **horizontal ground loop** is usually more economical than a vertical loop to install. Instead of digging wells, a backhoe or excavator is used to dig several trenches that are usually up to 6 feet deep and typically between 400 and 600 feet long. Several loops are run either in parallel or series through the trenches and can be layered at different depths. The most economical time to install a horizontal loop is during the early construction phase of the home or building. At this time the trenches can be dug along with the building's foundation or basement by the excavation contractor and back-filled accordingly. Similar to vertical loops, the individual horizontal loops are joined together by a common header that is piped back to the home or building.

When choosing to install a horizontal ground loop, the decision needs to be made as to whether to install a series or parallel loop. Series loops follow a single path from the beginning to the end of the loop (Figure 10-14). These types of loops are typically limited to a maximum heat pump capacity of 2 tons, but are easier to install. The main advantage to choosing a series loop configuration is in removing any excess air that may become trapped within the loop. It is very important that all air be removed from any closed-loop configuration. Trapped air can result in corrosion of any metal component within the water loop. In addition, any air in the loop will cause blockages that will restrict or prevent water from freely flowing, resulting in a loss of heat transfer and reduction in overall heat pump performance.

Following is a list of advantages and disadvantages of installing series flow systems:

Advantages:
- Ease in removing trapped air
- Flow path is simplified
- Higher heat transfer per foot of pipe

Figure 10-14

A horizontal series, single-layer ground loop configuration.

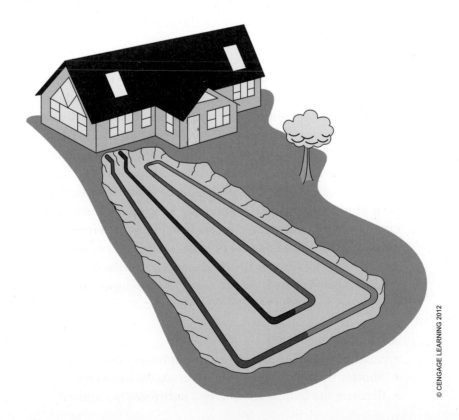

© CENGAGE LEARNING 2012

Figure 10-15

A two-layer series ground loop configuration. Notice that the piping is buried at depths of 4 and 6 feet.

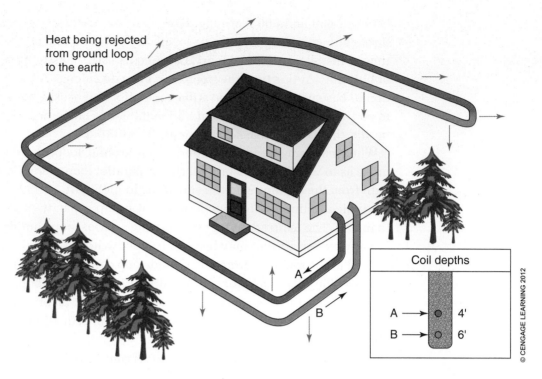

Heat being rejected from ground loop to the earth

Coil depths

A → 4'
B → 6'

Disadvantages:

- Larger-diameter pipe is required, resulting in more water/antifreeze solution being needed
- Higher excavating costs due to longer required trench length
- Higher pressure drop through the piping

Parallel-flow loop configurations are more popular due to the fact that a smaller-diameter pipe can be used (Figure 10-15). This can mean lower piping cost and reduced pressure drop through the loop. However, removing air from a parallel loop can become a problem due to the multiple paths that the water solution has to follow. The same problems with trapped air can be encountered in parallel-flow loops as with series loops. Each parallel path or circuit should contain equivalent lengths of piping to ensure there are equal pressure drops through each path. Otherwise, the path with the shortest length of pipe will receive the most flow, and the longest path will be starved. Heat pump performance is seriously affected by unequal flow though each path or circuit. Large-diameter headers are installed at both inlet and outlet locations near the building. These headers may be installed inside or outside of the structure depending upon preference (Figure 10-16). Headers are utilized to help to ensure equal flow and pressures between each path or circuit. Following is a list of advantages and disadvantages when installing parallel-flow systems:

Advantages:

- Smaller-diameter piping equates to lower cost.
- Shorter trenching results in lower excavating costs.
- Because the loop is shorter, less antifreeze is required.

Figure 10-16

A four-pipe parallel ground loop configuration. Notice that the piping is buried at depths from 3 to 6 feet.

© CENGAGE LEARNING 2012

Disadvantages:

- Air removal is much more difficult because of multiple circuits.
- Balancing water/antifreeze flow is a problem if circuits are of unequal lengths.

Slinky Loops

Figure 10-17 illustrates one of the most intriguing horizontal loop configurations—the **slinky loop**. This type of loop is also referred to as a coiled loop.

The slinky loop is a horizontal closed loop that consists of a continuous piping circuit that is overlaid in a circular fashion that resembles a slinky that has been uncoiled and flattened. One of the advantages of a slinky loop is that because of its compact design, the piping can concentrate the same amount of heat transfer into a smaller volume and can be installed into a shorter trench, thus saving on excavation costs. The coils of the slinky loop measure between 30 and 36 inches in diameter, with the spacing between each loop between 10 and 56 inches. This spacing is known as the loop's pitch. The pitch is determined by whether a compact or conventional loop is chosen. The compact slinky loop has a 10-inch pitch, which is equivalent to approximately 12 feet of pipe per foot of trench. By using a compact slinky loop, the user can reduce the trench length by about two-thirds compared to a conventional horizontal-parallel loop (Figure 10-18).

The specific required pitch and overall loop length for a given trench will vary depending on the climate and soil conditions for a given geographical area. Once the loop is laid out in the trench, a single return pipe is all that is needed. In addition to reducing excavation costs, another advantage of slinky loops is they

Green Tip

Air Separators and Automatic Air Vents

When implementing a geothermal parallel loop configuration, it may be wise to install an air separator and automatic air vent on each of the individual circuits to help prevent problems with trapped air in the system.

As the name implies, an air separator is used to separate air from the water as it flows through the system. The latest generation of air separators incorporates a mesh screen that causes the air to collide and adhere to it. As more air bubbles adhere to the mesh, they get larger, break loose, and travel up into the air vent (Figure A).

Automatic air vents are usually mounted on top of air separators and contain an internal disc that swells when water comes in contact with the vent. This swelling seals off the air vent port. However, as air accumulates around the disc, it becomes dry and shrinks, allowing the air to pass through the vent port. As the vented air is replaced by water, the disc once again swells, closing off the port (Figure B).

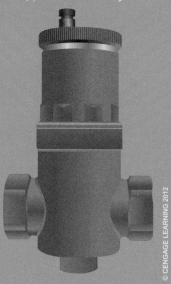

© CENGAGE LEARNING 2012

Figure A

An example of a wire screen air separator.

COURTESY OF DONALD STEEBY

Figure B

An automatic air vent.

may be installed on end in a narrow trench rather than laid flat. When doing this, it is very important that care be taken when back-filling the vertical trench due to possible damage to the piping and to ensure that no air pockets are left between the piping and the contact that it makes with the soil. Similar to the horizontal parallel loop, ¾-inch polyethylene piping may be used for a slinky loop.

Pond Loops

When the residential or commercial building is located near a body of water such as a pond or lake, a geothermal **pond loop** can offer a very economical alternative to other types of closed-loop systems (Figure 10-19).

There are several considerations when choosing this type of loop. The pond or lake should be at least 8 feet deep and the piping must be submerged to a depth that will prevent freezing conditions around the loop, even though the piping is filled with an antifreeze solution. Most loops are submerged to a depth that is within 1 foot of the bottom of the pond. Because the loop piping will float—even when filled with antifreeze—it must be weighted down. One method is to use a chain link fence that is placed over the entire pond loop. This will protect the coils and provide uniform weight over the entire length of the loop. It should be noted that the body of water is not adversely affected in any way by the pond loop. During the initial setup of the pond loop, the polyethylene piping is coiled and stacked into several individual loops, then floated out onto the pond, where it is submerged and weighted down. The supply and return header piping must be buried below the frost line according to the geographical region where the loop is being installed. Figure 10-20 illustrates an installation of a geothermal pond loop.

Figure 10-17

A compact slinky loop.

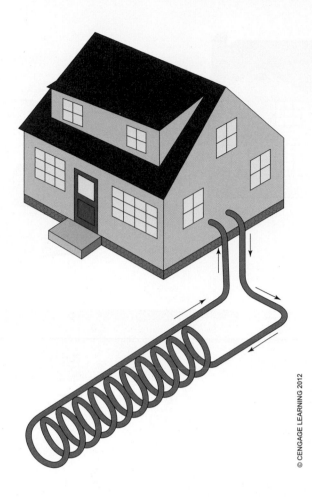

© CENGAGE LEARNING 2012

Figure 10-18

An installation of a slinky loop.

COURTESY OF STRANDLUND REFRIGERATION HEATING & COOLING

Figure 10-19

A pond loop.

COURTESY OF STRANDLUND REFRIGERATION HEATING & COOLING

Figure 10-20

An installation of a pond loop.

COURTESY OF STRANDLUND REFRIGERATION HEATING & COOLING

Domestic Hot Water Loop

Another type of geothermal heat pump loop is utilized for the heating of domestic hot water. This loop is known as a de-superheating loop and is piped from the hot gas discharge line of the heat pump's compressor to a separate heat exchanger. This heat exchanger is also a coaxial type, tube-in-tube heat exchanger similar to the one that is found in the heat pump cabinet (Figure 10-21).

Domestic water is pumped through the inside tube of the coaxial heat exchanger, where it is heated by the hot refrigerant gas that is passing in the opposite direction through the outside tube. As the domestic supply water is being heated as it passes through the heat exchanger, it is simultaneously de-superheating the hot refrigerant gas. In most cases, this configuration can supply up to 100% of the domestic hot water needs for most structures (Figure 10-22).

Figure 10-21

Separate connection ports on the heat pump for domestic water heating.

Figure 10-22

A domestic hot water loop showing the coaxial counterflow heat exchanger. The water is heated by the refrigeration loop.

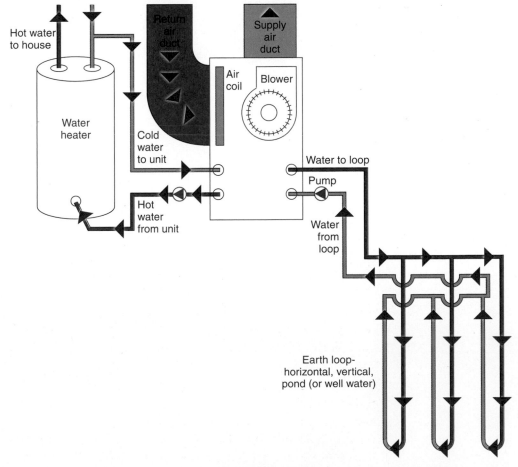

Chapter

11

PROPER SIZING PROCEDURES FOR GEOTHERMAL APPLICATIONS

When planning a successful geothermal project, it is important that the proper procedures be followed. Many times, a geothermal system simply does not perform well after it has been installed because the designer failed to use prudent sizing practices. In order to achieve successful geothermal heat pump applications and installations, designers should learn to master the following areas of expertise:

- Understand the proper procedure to develop accurate load calculations.
- Know what practices are used in proper duct sizing and installation.
- Utilize information to correctly select the proper equipment.
- Have a working knowledge of how the ground loop sizing is properly calculated.

LOAD CALCULATIONS

The first step in designing a successful geothermal system is to perform an accurate **load calculation** on the residential or commercial structure. An accurate load calculation will determine the structure's heat loss and heat gain based on the geographic location, the type of building materials used, and how well the building is resistant to the infiltration of unwanted and uncontrolled outdoor air. An accurate load calculation will accomplish the following goals:

- Ensure that the heat pump to be used will be properly sized
- Ensure that the duct system will be designed to deliver the correct amount of air to each room
- Ensure that the length and size of the ground loop are correct
- Provide the most comfortable conditioned space at the lowest operating cost

Residential and commercial load calculations are divided into two categories: **heat loss calculations** for determining heating requirements, and **heat gain calculations** for determining cooling requirement.

HEAT LOSS CALCULATIONS

Heat will migrate from a warm to a cold area. Because of this naturally occurring phenomenon, buildings will lose heat whenever the outdoor temperature is colder than the indoor temperature, no matter how well they are insulated. Therefore, it is easy to see why structures get cold in the wintertime (Figure 11-1).

Figure 11-1

Various heat loss characteristics of a given building.

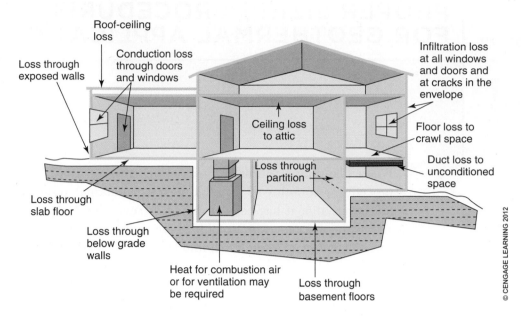

© CENGAGE LEARNING 2012

A typical procedure for calculating the heat loss of a given structure is as follows:

1. Select the proper outdoor winter design temperature based upon local conditions and meteorological data (Table 11-1).
2. Select the desired indoor design temperature to be maintained during winter conditions.

Table 11-1

Various U.S. Cities Showing Design Temperatures for both Summer and Winter

Location	Elevation Feet	Latitude Degrees North	Winter Heating 99% Dry Bulb	Cooling 1% Dry Bulb	Coincident Wet Bulb	Summer Design Grains 55% RH	Design Grains 50% RH	Design Grains 45% RH	Dally Range (DR)
Alabama									
Alexander City	686	33	22	93	76	39	46	52	M
Anniston AP	612	33	24	93	76	39	46	52	M
Auburn	776	32	22	93	76	39	46	52	M
Birmingham AP	644	33	23	92	75	34	41	47	M
Decatur	592	34	16	93	74	27	34	40	M
Dothan AP	401	31	32	93	76	39	46	52	M
Florence AP	581	34	21	94	75	31	38	44	M
Gadsden	569	34	20	94	75	31	38	44	M
Huntsville AP	629	34	20	92	74	28	35	41	M
Mobile AP	218	30	30	92	76	41	48	54	M
Mobile CO	26	30	29	93	77	46	53	59	M
Montgomery AP	221	32	27	93	76	39	46	52	M
Ozark, Fort Rucker	356	31	31	94	77	44	51	57	M
Selma-Craig AFB	166	32	26	95	77	42	49	55	M
Talladega	528	33	22	94	76	37	44	50	M
Tuscaloosa AP	170	33	24	94	77	44	51	57	M

Location	Elevation Feet	Latitude Degrees North	Winter Heating 99% Dry Bulb	Cooling 1% Dry Bulb	Coincident Wet Bulb	Summer Design Grains 55% RH	Design Grains 50% RH	Design Grains 45% RH	Dally Range (DR)
Alaska									
Adak, NAS	19	52	23	57	53	−18	−11	−5	L
Anchorage IAP	144	61	−9	68	57	−20	−13	−7	L
Anchorage, Elemendorf AFB	212	61	−8	69	57	−21	−14	−8	L
Anchorage, Fort Richardson	342	61	−13	71	58	−20	−13	−7	M
Annette	110	55	17	70	59	−14	−7	−1	L
Barrow	44	71	36	52	49	−25	−18	−12	L
Bethel	123	61	−24	68	57	−20	−13	−7	M
Betties	643	67	−44	75	59	−22	−15	−9	M
Big Delta, Ft. Greely	1277	64	−39	75	58	−27	−20	−14	M
Cold Bay	98	55	10	57	53	−18	−11	−5	L
Cordova	42	60	1	67	57	−18	−11	−5	M
Deadhorse	61	70	−34	61	54	−21	−14	−8	M
Dillingham	86	59	−13	66	56	−21	−14	−8	M
Fairbanks IAP	434	64	−41	77	59	−26	−19	−13	M
Fairbanks, Eietson AFB	545	64	−31	78	60	−23	−16	−10	M
Galena	152	64	−31	74	59	−21	−14	−8	M
Gulkana	1579	62	−39	73	56	−32	−25	−19	M
Homer	78	59	4	62	55	−18	−11	−5	L
Juneau IAP	19	58	7	69	58	−17	−10	−4	L
Kenai	92	60	−14	65	55	−23	−16	−10	M
Katchikan IAP	88	55	20	68	59	−11	−4	2	L
King Salmon	57	58	−19	67	56	−22	−15	−9	M
Kodiak	73	57	12	65	56	−19	−12	−6	L
Kotzebue	11	66	−31	64	58	−9	−2	4	L
McGrath	337	62	−42	73	58	−23	−16	−10	M
Middleton Island	87	59	21	60	51	−31	−24	−18	L
Nenana	362	64	−44	76	59	−24	−17	−11	M
Nome AP	37	64	−26	65	55	−23	−16	−10	L
Northway	1716	62	−32	74	57	−29	−22	−16	M
Port Heiden	105	56	−2	61	52	−29	−22	−16	L
Saint Paul Island	63	57	3	52	50	−22	−15	−9	L
Sitka	21	57	21	64	58	−9	−2	4	L
Talkeetna	358	62	−21	73	58	−23	−16	−10	M
Valdez	120	61	7	66	55	−25	−18	−12	L
Yakutat	33	59	2	63	55	−20	−13	−7	L

3. Determine the R-values for the construction materials of the building.
4. Calculate heat-transfer rates (BTU losses) on the building's envelope.
5. Calculate losses due to infiltration of cold outside air through cracks around doors and windows.
6. Calculate heat loads from ventilation requirements, and due to ducts located in unconditioned spaces.

This heat transfer between the warm indoor air and the cold outdoor air occurs primarily by two methods, conduction and convection. **Conduction** occurs when heat energy travels through the building materials. Naturally, if the building materials are of greater mass, the heat will tend to transfer at a slower rate. The amount of mass is referred to as the resistance to heat transfer and is expressed as the building material's R-factor. **Convection** is the transfer of heat energy through a gas or liquid. In any given building, this transfer occurs as a result of the amount of uncontrolled cold air that enters into and exits the structure.

The primary types of heat loss factors that influence residential and commercial buildings are:

- Transmission losses
- Infiltration losses
- Ventilation losses
- Duct losses

Transmission Losses

Transmission losses are due to the thermal transfer of heat energy through the building's envelope. These losses are a result of heat conduction naturally occurring between the indoors and outdoors. As mentioned earlier, the resistance of the building's materials to this heat transfer is referred to as the material's R-value. (Table 11-2 provides the R-values of various materials.) However, using the R-value in the heat-transfer formula can be clumsy and prone to error. Therefore, it is more common to utilize the material's U-value when calculating heating loads. The **U-value** of a material is defined as the quantity of heat in BTUs per hour (BTU/hr) that will flow through 1 square foot of material in 1 hour at a temperature difference between the indoors and outdoors of 1 degree. The U-value of a given building material is the reciprocal of the material's R-value. For instance, the U-value of a building material that has a value of R-5 would be 0.2 (1/5 = 0.2). A building material with a higher R-value results in a lower amount of heat loss, and so too does the material with a lower U-value.

Table 11-2

A Listing of Various Building Materials Showing their R-Values

Material		Thickness (Inch)	Density Lb/CuFT	R — Per Inch	R — As Listed
R-Values of Common Building Materials					
(Degrees F × SqFt)/(BTU/hr)					
4) Building Board, Sheathing, Panels — Continued					
c) Plywood		0.250	34	~	0.31
		0.375			0.47
		0.500			0.62
		0.625			0.77
		0.750			0.93
d) Plywood or Soft Wood Panel		1.00	34	1.25	~
e) Particle Board	Low Density	1.00	37	1.41	~
	Medium Density		50	1.06	
	High Density		62	0.85	
	Underlayment	0.625	40	~	0.82
f) Hard Board	Medium Density	1.00	50	0.73	~
	High Density		55	0.82	
			63	1.00	
g) Water Board		1.00	37	1.59	~
h) Fiber Board	Sheathing	0.500	18	~	1.32
		0.781			2.06
		0.500	22		1.09
		0.500	25		1.06
	Shingle Baker	0.375	18		0.94
		0.313			0.78
	Sound Board	0.500	15		1.35
h) Laminated or Pulped Paper Board		0.500	30	~	0.30
5) Membrane Material					
a) Permeable Felt		~	~	~	0.06
b) Two Layers Mopped 15 Lb Felt		~	~	~	0.12
c) Plastic Film		~	~	~	Negligible
6) Flooring materials					
a) Wood Sub Floor		0.750	~	~	0.94
b) Carpet and Fibrous Pad		~	~	~	2.08
c) Carpet and Rubber Pad		~	~	~	1.23
e) Cork Tile		0.125	~	~	0.28
f) Floor Tile or Linoleum		0.125	~	~	0.05
g) Terrazzo		1.000	~	0.08	~
h) Hardwood Flooring		0.75	~	~	0.68
7) Celling and Roof Deck Materials					
a) Fiberboard Tile and Lay-in Panels		0.50	18	~	1.25
		0.75			1.89
		1.00		2.50	~

Once the U-value is known, it can be used in the heat-transfer formula:

$$Q = U \times A \times TD$$

Where:

Q = heat loss in BTU/hr

A = area of the building surface in square feet

TD = temperature difference between the design indoor and outdoor in degrees Fahrenheit

This formula is used to determine the heat loss for each area of the structure, including walls, doors, windows, roof, and floor, and is typically inputted onto a spreadsheet.

The Air Conditioning Contractors of America (ACCA) publishes data and tables on R-values and U-values for various types of building materials. This information is available in the following publications: Manual J for residential load calculations and Manual N for commercial applications. These manuals will also combine the effects of U-values and design temperature differences into factors known as Heat Transmission Multipliers (HTMs). These values are also listed in various tables and can help reduce the vast amount of data that will need to be gathered when performing accurate load calculations.

Extensive information on R-values and U-values is also available from the American Society of Heating, Air Conditioning and Refrigeration Engineers (ASHRAE). This work is published as the *ASHRAE Handbook of Fundamentals* and includes extensive data on transmission values for various types of building materials as well as requirements for the correct amount of residential ventilation.

Infiltration Losses

Infiltration loss is a result of unwanted air that migrates into the building through cracks around doors and windows. Even though a structure may be classified as "air tight," infiltration occurs every time an outside door is opened. Losses from infiltration can be a substantial percentage of the total heat loss of the structure. This is why it is important to carefully measure this variable to ensure it is accurate. The formula used to calculate the amount of heat loss as a result of infiltration is:

$$Q = CFM \times 1.1 \times TD$$

Where:

Q = heat loss in BTU/hr

CFM = cubic feet per minute of infiltration from outdoor air

1.1 = a constant that converts the specific heat and density of standard air into BTU per degree Fahrenheit per hour

TD = temperature difference between the design indoors and outdoors in degrees Fahrenheit

The variable that needs to be determined in this formula is **cubic feet per minute**, or **CFM**. Traditionally there are two methods of calculating the CFM of unwanted infiltration air. The first is known as the air change method. This method is used to estimate the number of times the volume of air contained in the structure is displaced every hour. Tables are available to assist in estimating this factor. An example of a very-well-constructed or "tight" building would be 0.1 air changes per hour. Conversely, a structure that is "very leaky" may have up to 1.5 air changes per hour. Once this factor is determined, the CFM can be calculated by multiplying the total volume of the space above grade by the air change per hour, and dividing this number by 60.

The second method for calculating infiltration is the "crack" method. This method determines the linear feet of crack around doors and windows and is multiplied by the estimated CFM per foot of crack. Just as with the air change method, tables are available in order to determine these factors depending upon the types of doors and windows being used. Once the CFM is calculated, this variable is factored into the infiltration heat loss formula.

There is a third method that is used to determine the heat loss due to infiltration, the blower door test (Figures 11-2 and 11-3). This is a well-accepted diagnostic tool that incorporates a special exhaust fan and an air pressure gauge that is used to determine the pressure drop of the building over a fixed time. The results of this testing procedure are then used to determine the infiltration CFM of the structure.

Figure 11-2

Equipment used to perform a blower door test.

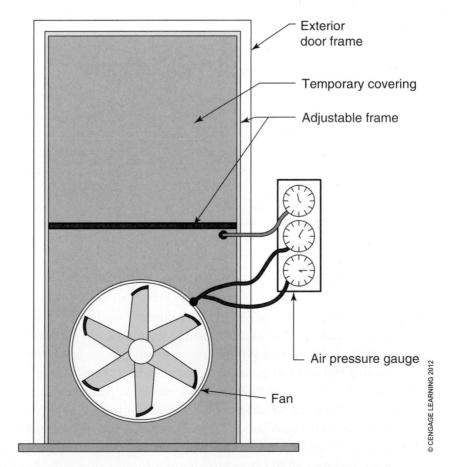

Exterior door frame

Temporary covering

Adjustable frame

Air pressure gauge

Fan

© CENGAGE LEARNING 2012

Figure 11-3

How a blower door test works.

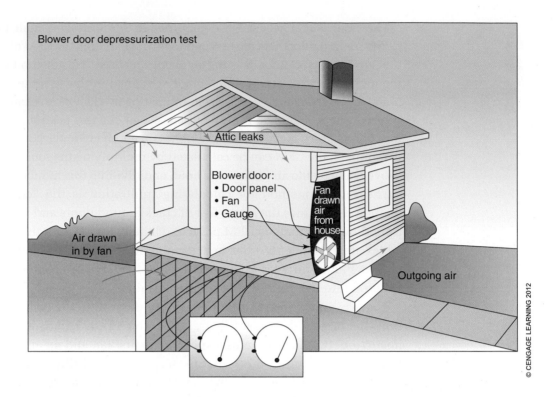

Blower door depressurization test

Attic leaks

Blower door:
• Door panel
• Fan
• Gauge

Fan drawn air from house

Air drawn in by fan

Outgoing air

© CENGAGE LEARNING 2012

Tech Tip
Considerations When Calculating Infiltration

When calculating heat losses from infiltration, remember that numerous factors can contribute to infiltration besides cracks around doors and windows. These include such things as: fireplaces, flue vents with no dampers on furnaces and water heaters, vents on bathroom and kitchen hood exhaust fans, and ventilation or make-up air ducts that may be connected to the return duct on the furnace.

Ventilation Losses

Ventilation losses are classified as outdoor air that is introduced into the conditioned space by mechanical means. Examples of ventilation air are exhaust fans in bathrooms and on kitchen range hoods. These factors are calculated by determining the CFM of each exhaust fan within the building and incorporating this number into the infiltration loss formula. Typical ventilation rates for bathroom exhaust fans are 50 CFM each, and kitchen hoods are typically rated at 100 CFM each. Because these fans are only operational on an intermittent basis, they do not play an important role in calculating heat losses, but still must be factored into the total heat loss on the structure in order to gain an accurate calculation.

Duct Losses

Heat losses from ductwork are typically categorized as either leaks from poorly sealed joints or transmission losses as a result of the duct running though an unconditioned space. A leaking duct may not play an important role in total structural heat loss if the duct is routed between floors that are heated. The reason being is that the air leaking from the duct is indirectly heating the conditioned space; therefore it is still being utilized. However, duct joints in commercial buildings should always be tightly sealed to prevent leakage that could significantly impact heat loss calculations.

Ductwork that is located in an unconditioned space must be factored into transmission losses, regardless of whether the structure is commercial or residential, and depends on the temperature difference between the air in the duct and in the unconditioned space. For this reason, this ductwork should always be thoroughly insulated.

HEAT GAIN CALCULATION

During the summertime, when the building is struggling to keep cool, warmer outdoor air is constantly trying to make its way inside. Therefore, calculating the heat gains of a building is done for exactly the opposite reason that heat losses are calculated. However, calculating heat gains is more complicated due to the fact that there are more factors involved (Figure 11-4). The same factors that affect heat loss (transmission, infiltration, ventilation) also have an effect on the heat gain of the building.

Figure 11-4

Various heat gain characteristics of a given building.

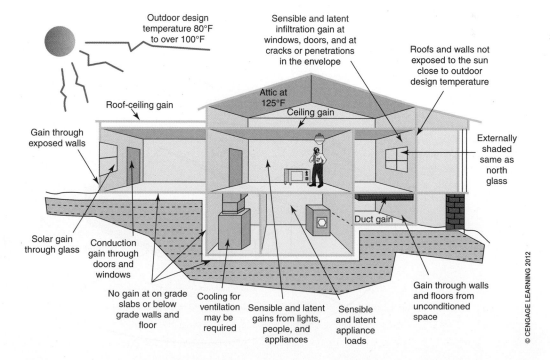

© CENGAGE LEARNING 2012

In addition, the following elements need to be determined when conducting an accurate cooling load calculation:

- The sun's radiation through roofs, walls, and windows
- Internal gains from people, appliances, and lighting
- Moisture that must be removed from the conditioned space

Calculating Solar Gain

Solar gain is defined as radiation from the sun shining on the roof, on the walls, and through glass. The intensity of this radiation is determined by several factors. These factors include:

- The orientation of the building in relation to its compass direction
- The color of the building
- The amount of window shading
- The time of year
- The time of day

The building's orientation has an effect on solar gain due to the fact that there is less radiation from the sun on the northern exposure versus the southern exposure (Figure 11-5). Therefore, expect higher solar gains on windows facing south. The color of the building also affects the amount of heat gain due to the fact that darker colors tend to absorb heat, whereas lighters colors will reflect heat. Again, expect higher solar gains on darker-colored buildings.

Window shading can have a significant effect on the reduction of solar gain. These shadings include such things as indoor window blinds and draperies, and awnings on the outside of the building.

Figure 11-5

The building's orientation has an effect on solar gain due to the fact that there is less radiation from the sun on the northern exposure versus the southern exposure.

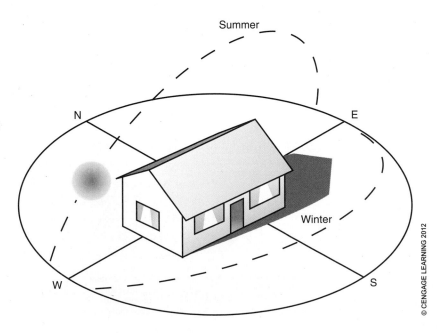

© CENGAGE LEARNING 2012

The angle of the sun's rays has a large impact on solar gain and will vary according to the season of the year. Because the sun is positioned higher in the sky during the summer months than in winter, solar radiation gains and their influence upon the cooling load will be the greatest in the months of July and August. From a time-of-day perspective, again because of the angle of the sun on the structure's glass, expect solar gains to be the greatest during the hours of 10 a.m. and 2 p.m.

Calculating Internal Gains

Another substantial factor when calculating the building's heat load is the effect of **internal heat gains** from people, appliances, and lighting. These internal gains are especially important to take into consideration when performing load calculations on commercial buildings. Occupants within the building will each generate approximately 300 BTU/hr when they are sedentary. This figure is much higher when there is strenuous activity, such as at a fitness center or bowling alley.

Internal gains from appliances can vary between residential and commercial applications. The kitchen is generally the location of the largest internal gains within a residential application, and a value of 1,200 BTU/hr is typically assigned to cover this area. However, in commercial applications, equipment such as computers, copiers, fax machines, coffee makers, and so forth will need to be taken into consideration in order to develop an accurate load calculation.

In addition to appliances and equipment, lighting plays an important role in calculating internal heat gains, especially in commercial applications. Although great strides have been made to reduce the power consumption of modern-day light fixtures, these devices still produce heat. The formula for calculating the BTU heat gains from appliances, equipment, and lighting is to multiply the wattage by 2.31, where 1 watt of electrical power is equal to 2.31 BTU/hr.

Calculating Latent Heat Gains

Latent heat can be considered the amount of humidity in the air. When calculating the cooling load for a building, the **latent heat gain** can be a substantial amount based on the geographic location, the number of people occupying the building, and the type of activity taking place in the building. In fact, the latent load can sometimes represent up to one-third of the total cooling load. In the air conditioning process, latent heat is removed from the structure by condensing the water vapor out of the air on the evaporator coil. This process does not change the temperature of the air, so latent heat cannot be measured conventionally with a thermometer, like sensible heat can. Rather, it is measured in BTU/hr. At 70°F, approximately 1,070 BTUs of heat energy are required to convert 1 pound of water to vapor. This process is known as the latent heat of vaporization and is the basis for dehumidification. Calculating the latent load of the building is extremely important due to the fact that problems with mold and mildew can be created if a system with a low latent cooling capacity is installed, especially in geographic areas with high humidity.

DUCT SIZING

Once the proper structural loads are calculated, the information obtained can be utilized to properly size the ductwork that will be used to deliver and circulate the airflow throughout the geothermal system (Figure 11-6). Close attention should be paid to this phase of the project for several reasons. Most conventional residential ductwork systems are undersized and are not designed to deliver the proper airflow needed for the modern geothermal system to operate properly.

Figure 11-6

Close attention must be paid to the geothermal duct installation to ensure it is sized correctly.

© ISTOCKPHOTO/CHRIS KRYZANEK

This is due to the fact that when a geothermal system is in the heating mode, the temperature of air delivered to the conditioned space is typically lower than that delivered by a conventional furnace. Because the supply air temperature is lower, the air velocity is typically slower than that of a conventional system as well. For this reason, the ductwork used with geothermal systems is typically larger than that of a typical residential or commercial application.

Tech Tip

Sizing Ductwork for Geothermal Applications

Because geothermal units deliver conditioned air to the space at lower temperatures than most conventional heating systems, two words should be remembered when sizing ductwork: *low* and *slow*. This means that because the supply air temperature is lower, the velocity of the supply air should be slower than normal to avoid possible draftiness. To compensate for this slower velocity of air, the supply fan should be sized properly as well to ensure the correct volume of air is delivered to the conditioned space.

Just as with load calculations, the ACCA publishes manuals for sizing ductwork for both residential and commercial applications. Manual D is used for residential sizing and Manual Q is typically used for commercial duct sizing.

Following is the recommended procedure for designing the duct system:

1. Select the type of air distribution system.
2. Calculate CFM values for each space based upon load calculations.
3. Select and locate air distribution devices.
4. Size ductwork based on velocity requirements.
5. Calculate system pressure losses.

Air Distribution Systems

The four most common duct configurations used in residential and commercial applications are radial duct, extended plenum, reducing extended plenum, and perimeter loop. Radial duct systems are mostly used in smaller, one-story structures and can be located in a crawl space or attic. With this system, all individual runs originate at the central plenum and are generally short in length. A single return air duct can be utilized with the radial duct system, making the installation more economical.

Extended plenum systems are applied to long structures such as ranch-style homes. Sometimes referred to as a truck duct system, the extended plenum can

be round, square, or rectangular. Smaller ducts called branches are connected to the main duct and feed individual spaces. One drawback in choosing the conventional extended plenum system is that the air velocity is reduced as it reaches the end of the trunk, possibly affecting the heating and cooling performance of the last space.

The reducing extended plenum system is most commonly used in selecting ductwork. One reason is that less material is used with this type of system, which reduces installation costs. Another is that because of its shape, the air pressure remains constant throughout the ductwork, which equates to better performance and air distribution throughout the system. Typically, the main duct is reduced every third of the way down the duct, or after every fourth branch take-off.

The perimeter loop system is best suited for slab-on-grade applications. The duct loop is installed under the concrete slab close to the outer walls of the building, with the outlets next to the wall. When the unit is in the heating mode, the entire slab is warmed, making it a well-suited system for single-story structures located in cold climates. Figure 11-7 shows examples of various ductwork systems.

Figure 11-7

Examples of various ductwork systems.

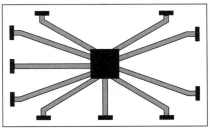

(A) Plenum or radial duct system

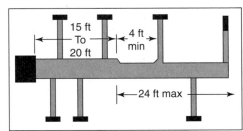

(C) Reducing extended plenum system

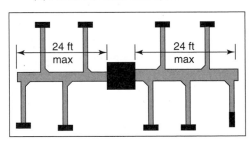

(B) Extended plenum system

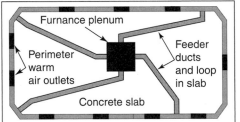

(D) Perimeter loop system with feeder and loop ducts in concrete slab

Calculating CFM Values

In order to achieve even temperature throughout the structure, airflow for individual spaces and branch ducts should be calculated based upon the heating and cooling load calculations. As stated earlier, the sensible heat formula is as follows:

$$Q = CFM \times 1.1 \times TD$$

Where:

Q = heat loss in BTU/hr

CFM = cubic feet per minute of infiltration from outdoor air

1.1 = a constant that converts the specific heat and density of standard air into BTU per degree Fahrenheit per hour

TD = temperature difference between the design indoors and outdoors in degrees Fahrenheit

Once the BTU/hr is determined for each space, the CFM requirement can be calculated by re-ordering the formula:

$$CFM = BTU / TD \times 1.1$$

Once this calculation is performed for both the heating and cooling loads, the larger of the two airflow figures should be selected.

Select and Locate Air Distribution Devices

For optimum airflow patterns throughout the structure, selection and placement of supply and return termination devices is critical. Typically, supply registers are placed at or near outside walls, preferably under windows (Figure 11-8).

This practice will create an air curtain along the wall that will reduce the transmission and radiation effect caused by windows. Return air grills are located on interior walls to create a more even air pattern between supply and return air throughout the space and reduce the level of stagnant air zones. Size supply and return terminal devices based on CFM requirements for the conditioned space. The ACCA has now made available Manual G for assistance in selecting the proper terminal devices for each space.

Figure 11-8

Supply registers located on an outside wall, under windows.

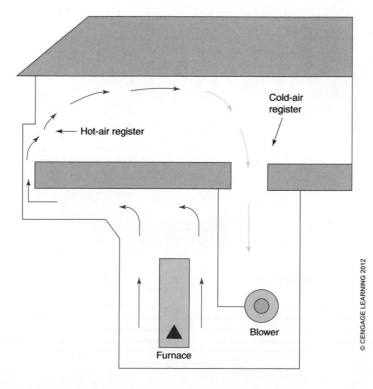

Cold-air register

Hot-air register

Blower

Furnace

© CENGAGE LEARNING 2012

Sizing Ductwork

The size of the ductwork selected should be based upon the air volume requirements, air velocity requirements, and static pressures for each trunk. Duct systems are classified as low-, medium-, and high-pressure systems, with low-pressure systems equating to lower velocity and lower static pressure. Static pressure is the pressure that is exerted on all areas of the duct when the system is pressurized by airflow. Velocity pressure is the pressure required to move the air up to its proper speed. ACCA Manual D recommends that a velocity of 900 feet per minute be maintained in the main trunk and 600 feet per minute in each branch. Do not exceed 700 feet per minute in each branch or noise will become a problem. The ACCA also recommends that at least one supply air outlet be provided for each 8,000 BTU/hr of room heat loss or 4,000 BTU/hr of sensible heat gain. Another method of cross checking airflow requirements is to provide between 400 and 450 CFM per ton of cooling for the entire structure. Each 6-inch round duct will typically deliver 100 CFM to the space, and each 8-inch round duct will deliver 200 CFM. The use of a duct calculator will also be an asset in properly sizing ductwork (Figure 11-9).

Figure 11-9

An example of a duct calculator.

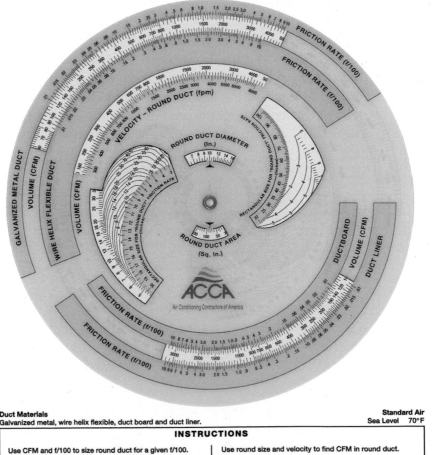

Duct Materials
Galvanized metal, wire helix flexible, duct board and duct liner.

Standard Air
Sea Level 70°F

INSTRUCTIONS

Use CFM and f/100 to size round duct for a given f/100.

Use CFM and velocity to size round duct for a given velocity.

Use CFM and round size to find velocity in round duct.

Use CFM and round size to find f/100 in round duct.

Use round size and velocity to find CFM in round duct.

Convert round size to a rectangular size with same f/100.

Convert diameter to round area.

See other side for rectangular duct areas and velocities.

Calculate System Pressure Losses

Air naturally wants to travel in a straight line; however, pressure losses occur whenever the system's airflow is required to change direction or pass through an obstruction. Losses also occur as a result of friction within the duct. These pressures are measured in inches of water column (in. W.C.).

Things to Know

MEASURING DUCT PRESSURE

Duct pressures are measured using such devices as a magenhelic gauge or U-tube manometer. These pressures are measured in inches of water column, or in. W.C.

Water column pressure is the amount of force required to raise a column of water. Inches of water column is used rather than pounds per square inch (psi) because the pressure in the duct is slight. As a comparison, 1 psi is equal to 27.7 in. W.C.

The typical ducted air delivery system consists of various fittings that cause the airflow to change direction. This directional change along with the friction caused by the air coming in contact with the ductwork creates static pressure losses that must be overcome. In addition to these issues, static pressure losses occur when the air passes through filters, heat exchangers, and evaporator coils. Ductwork losses can be calculated by the use of a friction chart. Pressure losses through fittings are determined by referencing the manufacturer's tables and charts or through Manual D. Also, the manufacturers of filters and evaporator coils have published listings showing static pressure losses through their equipment (Figure 11-10).

Tech Tip

Computer Programs for Duct Sizing

Just as with load calculations, there is computer software available to accurately size the air delivery system for most geothermal applications. Programs such as Ductsize by Elite Software (http://elitesoft.com) and Right-HV Duct (http://wrightsoft.com) can be used to ensure that duct sizing is done correctly and accurately.

Figure 11-10

Static pressure losses occur through various fittings in the duct work system.

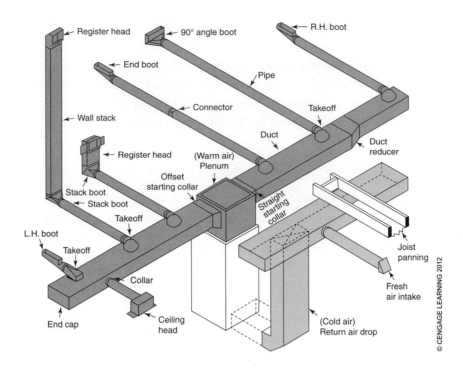

© CENGAGE LEARNING 2012

EQUIPMENT SELECTION

Along with duct sizing, a load analysis of the structure will be instrumental in choosing the correct heat pump for the project. The first decision to be made is whether the equipment should be selected based upon the heating load or the cooling load. In northern climates where there is a demand for heating and the cooling operation is minimal, it is a generally accepted practice to size the equipment for between 65% and 75% of the design heating load. In milder climates where the heating and cooling operational time is close to equal, size the heat pump for the exact heating load, but do not exceed the cooling load by more than 25%. In very warm climates where the cooling load is predominant, size the heat pump as close to the cooling load as possible. Regardless of whether the heat pump is selected for primarily heating or primarily cooling, do not intentionally oversize the equipment in an attempt to compensate for the unknown. Oversizing will cause the equipment to cycle off and on too frequently. This will result in poor performance, an uncomfortable conditioned space, and increased utility bills. Oversizing can also shorten the life of the equipment. One way to compensate for equipment selection that is near the optimum heating or cooling load is to choose a heat pump that is equipped with two-stage operation. This will allow the equipment to operate at partial capacity when the structure is not under a full load demand for either heating or cooling. For instance, if the heat pump is located in a cold climate and is chosen for the optimum heating load, a two-stage or modulating heat pump can operate at a lower stage during the cooling season without the risk of it being oversized and suffering from poor performance.

When checking the heat pump manufacturer's data for selection criteria, keep in mind that the heating and cooling capacities are based on the temperature of the water from the ground loop entering the coaxial heat exchanger. The heat pump should be selected for heating based upon the coldest entering water temperature (EWT), whereas cooling should be based on the warmest entering water temperature (see Table 11-3).

Table 11-3
Geothermal Heat Pump Data Including Performance Under Various Entering Water Temperature Conditions. Performance Data Courtesy of Water Furnace International

| | | | Flow Rate | | Water Loop Heat Pump | | | | Ground Water Heat Pump | | | | Ground Loop Heat Pump | | | |
| | | | | | Heating EWT 68°F | | Heating EWT 68°F | | Cooling EWT 59°F | | Heating EWT 50°F | | Cooling Brine Full Load 77°F Part Load 68°F | | Heating Brine Full Load 32°F Part Load 41°F | |
Model	Cap. Mod	Motor	gpm	cfm	Capacity BTU/hr	EER BTU/hr/W	Capacity BTU/hr	COP	Capacity BTU/hr	EER (BTU/hr)	Capacity BTU/hr	COP	Capacity BTU/hr	EER BTU/hr/W	Capacity BTU/hr	COP
026	Full	ECM	8	950	26,000	16.0	31,000	5.5	29,000	24.0	25,300	5.0	27,200	18.6	19,500	4.2
	Part	ECM	7	750	19,500	18.6	22,600	6.3	22,000	31.2	18,100	5.4	21,500	26.8	16,200	4.7
038	Full	ECM	9	1300	39,000	17.2	42,200	5.5	39,400	24.1	34,800	5.0	40,200	20.1	27,000	4.2
	Part	ECM	8	1150	28,000	20.1	30,300	6.5	30,500	32.1	24,800	5.4	30,100	30.0	22,300	5.1
049	Full	ECM	12	1400	48,300	15.8	57,400	5.1	53,200	22.7	47,200	4.7	50,000	18.0	37,400	4.1
	Part	ECM	11	1200	35,900	18.1	41,900	6.1	37,800	28.3	34,000	5.2	38,700	25.1	31,000	4.7
064	Full	ECM	16	1800	64,500	16.2	72,500	5.1	70,700	22.7	56,800	4.6	67,600	18.0	45,800	3.9
	Part	ECM	14	1500	47,000	18.2	51,500	5.8	51,500	29.3	39,600	4.8	51,100	25.6	36,000	4.2
072	Full	ECM	18	2000	71,000	15.0	86,700	5.0	79,900	20.4	67,900	4.4	73,600	16.8	54,100	3.8
	Part	ECM	16	1800	54,000	16.6	63,400	5.4	62,200	26.0	51,000	4.6	58,800	23.1	45,000	4.3
022	single	ECM	8	800	20,700	17.5	25,300	6.2	23,500	30.0	19,800	5.3	21,700	21.0	15,000	4.0
	single	PSC	8	750	20,600	17.2	25,000	6.0	23,000	28.0	19,800	5.0	21,200	20.3	15,000	3.8
030	single	ECM	8	1000	28,300	19.2	32,700	5.8	31,300	28.8	25,800	5.0	29,400	21.9	20,000	4.0
	single	PSC	8	900	28,100	18.2	32,700	5.5	30,900	27.1	25,800	4.8	29,200	21.1	19,800	3.8
036	single	ECM	9	1200	34,500	19.6	38,000	6.1	37,200	30.1	30,300	5.2	35,000	22.0	24,100	4.4
	single	PSC	9	1200	34,100	17.6	37,900	5.6	36,300	25.7	30,300	4.7	34,600	19.6	24,100	4.0
042	single	ECM	11	1300	40,600	19.2	44,100	5.9	45,200	29.5	34,900	5.2	42,000	21.4	27,500	4.2
	single	PSC	11	1300	40,100	16.6	44,100	5.3	44,600	24.5	34,900	4.6	41,600	18.6	27,500	3.7
048	single	ECM	12	1500	47,000	17.5	55,400	5.5	52,000	26.1	45,100	4.8	49,300	19.7	35,300	4.0
	single	PSC	12	1500	46,400	15.5	55,400	5.0	51,600	22.5	45,100	4.3	48,900	17.3	35,300	3.6
060	single	ECM	15	1800	64,300	17.2	69,800	5.4	72,000	26.1	55,100	4.7	66,800	19.5	43,200	3.9
	single	PSC	15	1800	64,000	16.0	69,800	5.1	71,700	24.6	55,100	4.4	66,800	18.5	43,200	3.7
070	single	ECM	18	200	70,600	16.0	84,300	5.1	79,100	23.8	66,100	4.4	73,200	18.2	52,000	3.7
	single	PSC	18	2000	70,600	15.1	84,300	4.7	77,500	21.6	66,100	4.0	73,200	17.2	52,000	3.4

Cooling capacities based upon 80.6°F DB, 66.2°F WB entering air temperature
Heating capacities based upon 68°F DB, 59°F WB entering air temperature
All ratings based upon 208V operation

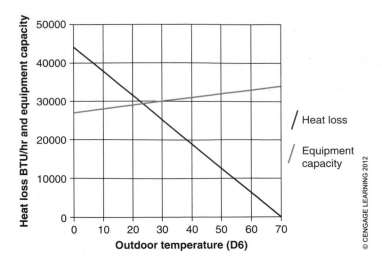

Figure 11-11

Equipment balance point. Note that this heat pump has a balance point of approximately 22°F. Auxiliary heat is sized to the left of the balance point.

Water temperatures will vary depending upon geographical location and soil type. These criteria should be used regardless of whether the system is utilizing either an open or closed loop.

Another equipment selection issue is the equipment's balance point. The balance point is the relationship between the system's capacity and the calculated heating load. The balance point occurs when the heat pump is delivering the same amount of heat to the space as is being lost. When the heat pump is performing above its balance point, it will cycle on and off. When the heat pump is operating below its balance point, it will run continuously. Most equipment that is selected does not have a balance point equal to the winter outdoor design temperature (Figure 11-11), which would require equipment that is too large and too costly. The solution to this issue is to include a form of auxiliary heat. Usually auxiliary heat is provided by the use of electrical strip heat, which is also typically the most expensive form of space heating. The reason why electric heat is used is because it is cheaper to install and maintain than heat systems using other conventional types of fossil fuels. Also, the length of time that the auxiliary heat is operating in the heating season is not substantial. Therefore the savings that are gained using a smaller-sized geothermal heat pump outweigh the additional cost of using electric heat.

Once the heat pump has been sized for the load requirements, check the CFM volume of the blower motor to ensure it is within range. One method of determining this is by calculating the system's nominal air volume. The nominal air volume is estimated by using the following equation:

$$CFM = \text{Sensible Load} \ / \ 1.1 \times TD$$

Where:
Sensible Load = calculated heating or cooling load from previous examples
$\qquad$ 1.1 = constant
$\qquad$ TD = temperature difference between supply and return air across the unit

The design temperature difference can be determined by deciding the design indoor air temperature (this will be the return air) and checking the manufacturer's supply air temperature data.

LOOP SIZING

Once it has been determined that a closed-loop geothermal system will be used over an open-type system, the final procedure will be sizing the ground loop. Regardless of whether a vertical, horizontal, slinky, or pond loop is chosen, certain conditions will preclude the length of the loop that will be required. The following conditions will determine the overall required loop length:

- The heat pump's capacities for heating and cooling
- The heat pump's performance data (Coefficient of Performance [COP] for heating, and Energy Efficiency Ratio [EER] for cooling)
- The type of piping used
- The local geological soil conditions
- The local geographic weather conditions for heating and cooling seasons
- The minimum and maximum design entering water temperature

The heat pump performance is determined by the manufacturer's equipment data. The type of piping to be used will determine the heat-transfer capability. Types of local soil conditions can be classified as follows:

- Dry, light soil
- Damp, light soil
- Dry, heavy soil
- Damp, heavy soil
- Wet soil or rock

Field Tip

Average Loop Lengths Based on Configuration

Here are some typical loop lengths depending on the type of configuration:

Vertical = 130 to 300 feet per ton, depending on soil conditions

Horizontal = 400 to 600 feet per ton

Slinky = 200 to 300 feet of trench per ton

Horizontal Parallel = provide one circuit per each ton of cooling

The type of soil in a given location can be determined by taking random soil samples where the loop is to be installed and having them analyzed. The local agricultural county extension service can be a resource for analyzing soil samples.

Once this information has been collected, it can be inputted into a complex mathematical formula to determine the proper loop length. Fortunately, as with load calculations and duct sizing, computer programs are available that will make sizing the loop length easier to do. Programs that are currently available include CLGS by Oklahoma State University (http://igshpa.okstate.edu), Right-Loop (http://wrightsoft.com), and ECA by Elite Software (http://elitesoft.com).

INSTALLATION AND START-UP OF GEOTHERMAL SYSTEMS

INSTALLATION PRACTICES

Once the proper heat pump model has been chosen and the ground loop is in place, it is important that the indoor equipment be correctly installed and commissioned. Proper installation and commissioning procedures will ensure that the heat pump operates at peak performance and that any required future service issues can be handled easily. This section will cover the proper steps that are necessary to execute a successful installation procedure.

Accessibility and Location

This includes the proper placement of the new equipment, usually in the basement of a home or in the mechanical room of a commercial building. The installer should always follow the manufacturer's installation instructions, which include having access to the equipment for ease of maintenance (Figure 12-1). There should be enough clearance to remove the heat pump's access panels for filter removal. Water and loop connections should be accessible as well. Ensure that the unit is slightly elevated off the floor in case there should ever be any accidental flooding.

Figure 12-1

A proper heat pump installation requires that it is accessible for service.

COURTESY OF DONALD STEEBY

Duct Connections

The ductwork should be checked to ensure that it is sized and installed properly (Figure 12-2).

This includes the proper sealing and insulating of any ducts that run through an attic or crawl space (Figure 12-3). Ensure there is adequate space for all duct transitions, return air drops, filter racks, and elbows. It is suggested that a flexible connection be used on the supply duct that feeds off the main **plenum** (Figure 12-4). This will reduce the chance of any unnecessary vibrations that may be magnified by the ductwork. As mentioned earlier, ensure that the ductwork is properly sized for adequate airflow. Large pieces of duct should always include a cross-break.

Figure 12-2

Be sure ductwork is securely fastened before starting up the heat pump.

COURTESY OF DONALD STEEBY

Figure 12-3

Sheet metal joints should be properly sealed to prevent air leakage.

COURTESY OF DONALD STEEBY

Figure 12-4

Flexible duct connections should be used to eliminate problems with vibration and noise.

ISTOCKPHOTO/LISA MCDONALD

Figure 12-5

The main disconnect should be mounted within 3 feet of the unit for ease of servicing.

COURTESY OF DONALD STEEBY

Electrical Connections

Install all line-voltage and low-voltage wiring according to local and national electric codes. Refer to the manufacturer's installation material to determine the proper wire size, proper fuse sizing, and proper grounding. The disconnect switch for the line-voltage connection should be located within 3 feet of the unit for convenience when servicing (Figure 12-5).

Be sure that the supply voltage reading is within plus or minus 10% of the data plate rating. The technician should have a good working knowledge on how to properly use a multimeter for measuring voltage, amperage, and for checking wiring continuity. Additional compliance considerations for wiring requirements can be found in the 2011 National Electric Code (NEC).

All low-voltage control wiring should be solid or stranded copper and have a minimum wire size of 18 gauge. Control wiring is typically bundled together in groups of four to eight wires, and the amount needed will depend on the number of accessories and features offered with the heat pump (Figure 12-6).

It is suggested that one or two extra wires be pulled as spares in case one wire breaks or other accessories are added to the heat pump at a later date. Avoid splicing control wiring, as this could lead to potential breakage that can jeopardize the continuity of the wiring. Some thermostats with solid-state electronics require the control wiring to be shielded cable. Consult the manufacturer's recommendations for proper control wiring.

Figure 12-6

Control wiring for the thermostat typically consists of four to eight wires bundled together.

COURTESY OF DONALD STEEBY

Drain Connections

The indoor coil on the heat pump should always include a condensate drain, and the drain should incorporate a trap (Figure 12-7). The trap prevents water from backing up into the coil, which results from the air velocity rushing past the drain hole on the condensate pan. The condensate drain line should be at least three-fourth inch PVC and should have a slope of at least one-fourth inch per foot toward the drain.

© CENGAGE LEARNING 2012

Green Tip

Proper Color Coding for Control Wiring

Technicians should memorize the standard color coding for the thermostat wiring to the heat pump. The following is the standardized color coding for heat pumps, furnaces, and air conditioners:

- **Red:** Common to the thermostat (This is the 24-volt supply voltage running from the heat pump to the thermostat.)
- **White:** Heating circuit (This may be the auxiliary heating circuit on the heat pump.)
- **Yellow:** Cooling circuit (This circuit typically energizes the compressor on the heat pump.)
- **Green:** Fan/blower circuit
- **Orange:** Reversing valve circuit (This circuit will be the switchover between heating and cooling.)

Other additional wires may be used for auxiliary devices such as a humidifier or outside air sensor. Always consult the manufacturer's literature to ensure the thermostat is wired correctly.

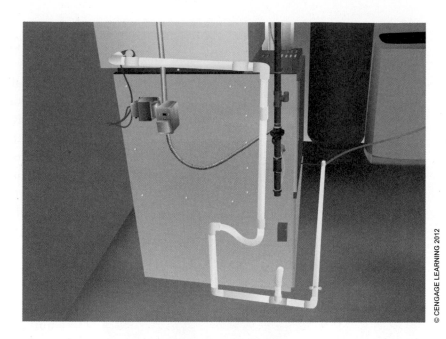

Figure 12-7
Indoor coils should always incorporate a drain line with a trap.

Heat pumps that are installed in areas where there is no easily accessible drain should include a condensate pump. This pump contains a sump and float switch that moves the water away from the drain pan and channels it to the nearest drain. Be sure to include a check valve when installing a condensate pump, especially if the condensate is being pumped vertically.

Ground Loop Connections

All ground loops should be made of polyethylene or polybutylene material. All joints buried in the ground should be joined by heat fusion. There should be no galvanized or steel fittings used with the ground loop piping, as these types of fittings are subject to corrosion. Avoid any plastic-to-metal threaded fittings that can result in potential leaks in earth-coupled applications. Ground loops that penetrate any exterior walls should be properly routed to the heat pump, and the holes sealed with a waterproof adhesive or hydraulic cement to prevent ground water intrusion (Figure 12-8).

Include **P/T plugs** on both inlet and outlet piping on the heat pump (Figure 12-9). These will be used to measure the pressure drop across the unit to determine the proper water flow. A conversion chart furnished by the manufacturer is used to read the water flow based on the pressure reading. Another practice is to install a water flowmeter instead of the P/T plugs (Figure 12-10). This device will make it convenient to monitor the water flow through the heat exchanger to ensure it is at the proper rate.

Figure 12-8

The ground loop that penetrates the exterior wall should be properly sealed and the wall insulated.

Figure 12-9

A P/T plug, or Pete's plug, should be installed on both the inlet and outlet ground loop water piping at the heat pump. This is used for taking pressure readings through the heat exchanger to determine water flow.

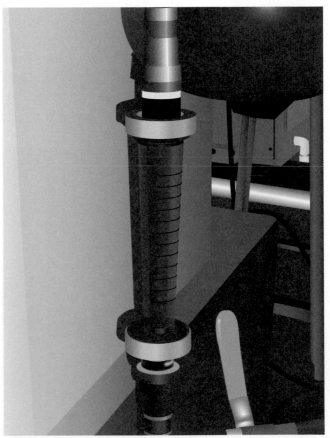

Figure 12-10

A flowmeter will make it easy to check the water flow through the heat exchanger.

Things to Know

WHAT IS A P/T PLUG?

P/T plugs, or Pete's plugs, are used for taking pressure and temperature readings across the ground loop water lines. The plugs are really valves consisting of a brass housing with ¼" pipe threads. A ⅛" needle, similar to the device used for pumping basketballs and footballs, is inserted into the plugs for taking pressure and temperature readings. The plugs also incorporate a self-closing valve for rapid sealing upon removal of the needle. The Pete's plug was developed by Mr. Charles D. Peterson in 1965.

Manual shut-off valves should also be included on both inlet and outlet piping for isolation purposes. These valves should be ball valves, and not gate valves (Figure 12-11). The seats on gate valves typically erode over time, leading to water leaking past the gate when they are shut off for servicing procedures. Before the heat pump is initially started up, the ground loop will need to be flushed and filled with antifreeze and water. Then the loop will need to be purged of any air and the proper flow adjusted at the pumping center. The use of a flush cart is sometimes used for filling and purging the ground loop.

COURTESY OF DONALD STEEBY

Figure 12-11

Ball valves rather than gate valves should be used as shut-offs. Gate valves typically erode over time, causing leaking past the gate.

System Airflow

Before the refrigeration side of the heat pump is energized, the system should be checked for proper airflow. This can be done by switching the thermostat to the fan-only setting and measuring the airflow using a manometer, magenhelic gauge (Figure 12-12), or velometer (Figure 12-13) to record the airflow velocity.

Once the velocity is known, the CFM, or cubic feet per minute air volume, can be calculated using the following formula:

$$\text{CFM} = \text{Velocity} \times \text{Area of the duct opening in square feet}$$

Airflow that is too high can create drafts and cause the occupants to feel uncomfortable. If the airflow is too low, the distribution efficiency falls and there can be accelerated wear on the system components, which can lead to premature failure.

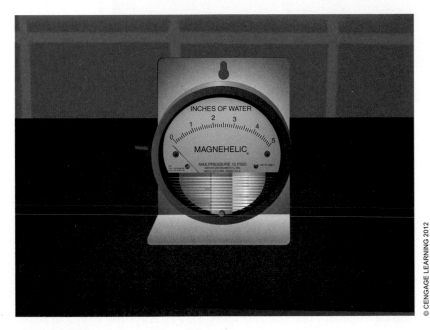

Figure 12-12

A magenhelic gauge used to measure airflow through the ductwork.

Figure 12-13

A velometer used to measure velocity through the duct.

Refrigeration Considerations

Last, the system should be checked to ensure the refrigerant charge is correct. An overcharged refrigerant system can be just as detrimental as one that is undercharged. Recent studies show that 7 of every 10 installed cooling systems have an incorrect refrigerant charge, and that some systems are overcharged by as much as 100%! An incorrect refrigerant level can lower the efficiency of the system by

Green Tip

R-22 versus R-410A Refrigerants

For years the refrigerant R-22 has been used in air conditioning systems and in heat pumps with great success. However, because of fears concerning greenhouse gas emissions and ozone depletion, R-22 has been banned from production in the United States as of January 1, 2010. To take its place, the refrigerant industry has developed R-410A. This replacement refrigerant has a higher BTU per pound value than R-22, making it more efficient for use in heat pumps; especially in colder regions of the United States. R-410A is now successfully used as a substitute for R-22; however, its success hasn't been without issues. R-410A has operating pressures that are up to 75% higher than R-22 pressures. For instance, to achieve a 40°F evaporator coil temperature, R-22 operates at a pressure of about 68 psi. However, the same coil temperature using R-410A equates to an operating

(Continued)

Figure 12-14

Refrigeration gauges should be connected to the heat pump only as a last resort to prevent refrigerant losses from the system.

as much as 5% to 20%. This issue can also cause premature equipment failure, resulting in costly repairs that could be prevented. Keep in mind that most geothermal heat pumps contain a critical charge of refrigerant, which means that attaching refrigeration gauges should be done as a last resort (Figure 12-14). Anytime a typical set of **manifold** gauges is attached and removed from the system, there is the potential to lose several ounces of refrigerant from the unit through the hoses.

It should be noted that the installation technician must be certified by the Environmental Protection Agency (EPA) to handle refrigerants. Section 608 of the Clean Air Act of 1990 outlines the proper service practices for recovery and recycling of ozone-depleting refrigerants. Section 608 also sets forth the standards for the proper equipment to be used for recovery and recycling, as well as for the proper disposal of refrigerants. Technicians are required to pass a written test in order to become EPA Section 608 certified.

Figure 12-15

Tanks of R-22 and R-410A refrigerant.

pressure of approximately 118 psi! To maintain safety when handling R-410A, technicians must be aware of these higher operating pressures and use the correct equipment and handling practices to avoid personal injury (Figure 12-15).

Green Tip (Contd)

START-UP PROCEDURES

When the heat pump has been installed, the next step is to perform a proper start-up procedure. This step is critical to the correct operation and long-term health of the heat pump. Following is a typical checklist that should be followed in the field to ensure a successful start-up procedure.

Before the unit is actually energized, be sure that the main electrical power has been applied to the unit for at least 24 hours to energize the crankcase heater. This step will ensure that there is no residual liquid refrigerant inside the compressor crankcase. Next, make a final visual inspection of the unit to ensure that the blower wheel spins freely, all electrical connections are tight, ductwork is properly connected, loop connections are secure, and access panels are in place (Figure 12-16).

Figure 12-16

Make final inspections to the system before doing an initial start-up.

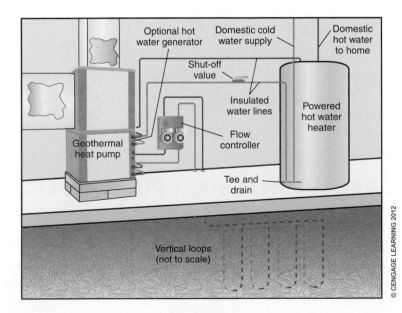

Optional hot water generator
Domestic cold water supply
Domestic hot water to home
Shut-off value
Insulated water lines
Powered hot water heater
Flow controller
Geothermal heat pump
Tee and drain
Vertical loops (not to scale)

© CENGAGE LEARNING 2012

Start-up Procedure Checklist

The following checklist should be followed when doing a start up procedure (Figure 12-17):

1. Turn on the main power supply.
2. Set the thermostat to the "manual fan" position.
3. Once the blower is energized, check for proper airflow through the diffusers.
4. Check the amp draw of the blower motor to ensure it is within specifications.
5. Set the thermostat to the "auto" position, so the blower will de-energize.
6. Set the thermostat to "heat" and move the setpoint above the room temperature.
7. Check to make sure that:
    a. The compressor is operating (there may be an on-delay timer that needs to time out).
    b. The blower is operating.
    c. Water is flowing through the heat exchanger.
8. Check the amperage of the compressor and compare it to the manufacturer's specifications.
9. Measure the water flow to see if it is within specifications.
10. Ensure that there is adequate temperature rise across the indoor coil.
11. Repeat this same process for the cooling mode.

Figure 12-17

A sample final equipment installation.

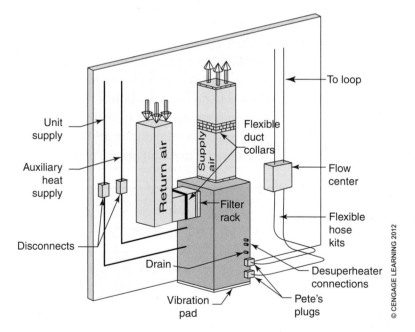

ADDITIONAL GEOTHERMAL HEAT PUMP APPLICATIONS

Up to this point, the focus has been on utilizing geothermal heat pumps for air-side applications. This is primarily known as a **water-to-air heat pump** system. However, there is another application for geothermal that is gaining popularity and can be used for several different purposes, as explained next.

COURTESY OF DON STEEBY

Figure 12-18

A water-to-water heat pump installed as part of a hydronic heating system.

Water-to-Water Heat Pumps

Water-to-water heat pumps utilize two coaxial heat exchangers instead of just one. They can be used on both open- and closed-loop systems (Figure 12-18).

Instead of using a hot water coil and blower to distribute warm air to the conditioned space, the second heat exchanger pumps tempered water to a network of floor-mounted piping, such as with a radiant floor heating system (Figure 12-19). Additional applications for water-to-water heat pumps include hot tubs, spas, swimming pools, and domestic water heating.

Figure 12-19

An example of a heat exchanger configuration on a water-to-water heat pump system.

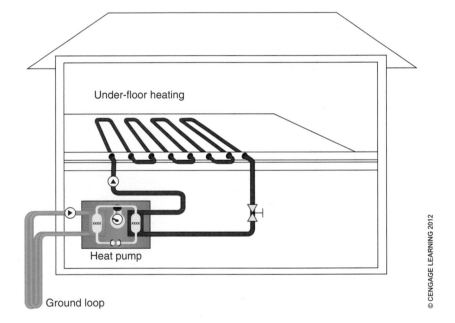

© CENGAGE LEARNING 2012

These are all viable options for delivering hot water at a reasonable cost. When the water-to-water heat pump is operating in the heating mode, the ground loop functions as the evaporator. The secondary heat exchanger acts as the condenser by heating the water.

When using a water-to-water heat pump on a multi-zone radiant floor heating system, there may be times when only one or two zones are calling for heat. In this situation, the water flow through the secondary heat exchanger will be restricted and can lead to higher-than-normal head pressures on the refrigeration side of the system, causing potential lock-out. To resolve this issue, it is common to incorporate the use of a buffer tank on the condenser water side of the system (Figure 12-20).

Figure 12-20

A buffer tank is located between the heat pump and the building heating devices.

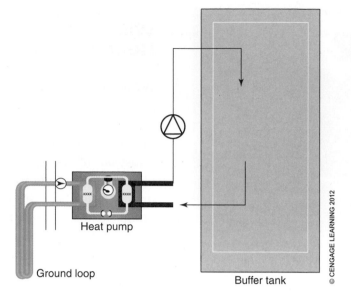

Heat pump

Ground loop

Buffer tank

© CENGAGE LEARNING 2012

Figure 12-21

The piping connections between the heat pump, buffer tank, and heating circuits.

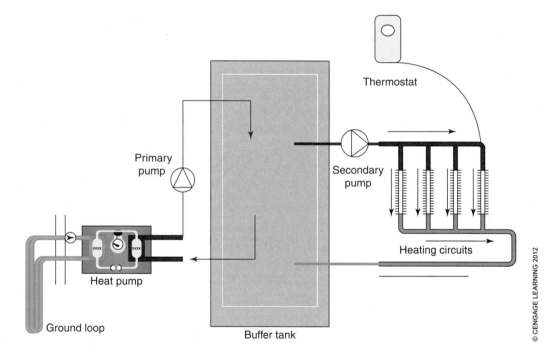

Thermostat

Primary pump

Secondary pump

Heating circuits

Heat pump

Ground loop

Buffer tank

© CENGAGE LEARNING 2012

The buffer tank acts as a storage vessel that receives water from the condenser and supplies tempered water to the hot water distribution system. With this configuration, the heat pump will cycle on and off to maintain a constant water temperature within the buffer tank. There are also two circulating pumps that are incorporated on the condenser side of the system. One of the pumps circulates the tempered water from the condenser to the tank. The other pump distributes the heated water from the buffer tank to the heating zones serving the conditioned spaces. This secondary pump can be controlled by individual zone thermostats (Figure 12-21).

When there is a call for heat from one of the zones, a zone valve for that space will open and the secondary pump will be energized. As more zones call for heat, the temperature of the water returning to the buffer tank will decrease, causing the heat pump to cycle on for longer periods of time to maintain the temperature setpoint within the buffer tank.

Case Study

—Gary and Lois Vanduine

Retrofit Success

As a means of combating the rising cost of heating their home with propane, Gary and Lois Vanduine of Middleville, Michigan, explored the possibility of installing a geothermal heat pump. Unfortunately, their existing system was a conventional hot water boiler, which made it seem impractical to invest in both a heat pump and a complete airside system. However, in 2003 they were convinced that it was worth the cost to install new ductwork and a new geothermal closed-loop heat pump.

By working through their local electrical cooperative, Great Lakes Energy, they chose an Econar Model GV36 heat pump with a capacity of approximately 36,000 BTU. They also decided that a slinky loop would be the most practical closed-loop configuration, and would take advantage of their existing backyard space (Figure 12-22).

Figure 12-22
Installing the slinky loop.

Figure 12-23
Additional electric meter dedicated to the heat pump.

(Continued)

Case Study (Continued)

The total package cost was $12,545, of which $3,600 was for the ground loop installation. As a money-saving incentive, Great Lakes Energy arranged to have a dedicated electric meter installed next to the existing meter that would be used exclusively for their heat pump (Figure 12-23).

This new meter would give them a 30% discount on their heat pump utility cost. In exchange, the Vanduines agreed to have interruptible service on the new meter. The dedicated meter also allowed Gary and Lois to track the money they would save using geothermal.

The initial estimated payback was about 6 years, but Gary says it was less than that. His average heating bills for January and February are about $60 each, which also happen to be the coldest months of the year in Michigan. The existing boiler is still operational (Figure 12-24). The Vanduines use it to warm the house occasionally or if it gets really cold outside. However, Gary and Lois agree they could probably do just fine without it.

COURTESY OF DONALD STEEBY

Figure 12-24

Heat pump installed next to hot water boiler.

UNIT 4

Biomass: Utilizing Wood, Corn, and Pellets as Heating Fuels

USING BIOMASS AS A HEATING SOURCE

13

INTRODUCTION

Heating with solid fuels has been keeping people warm for thousands of years. However, since the advent of the modern-day furnace, solid-fuel appliances such as wood stoves and wood-fired furnaces have become somewhat obsolete. Rather than toiling with cutting wood, stoking the wood stove, tending the fire, and removing the ashes from the bottom of the furnace, the general public has grown accustomed to simply adjusting a thermostat and walking away. The only wood-burning appliance that has remained in the home is the fireplace. However, its function has been more aesthetic in nature rather than a necessity to keep people warm.

This scenario has changed since the cost of energy has become a major part of the homeowner's and business owner's budget. With the ever-increasing cost of petroleum-based fuels such as coal, natural gas, fuel oil, and propane, it only makes sense to investigate alternative sources for producing heat. And what better source than one that is renewable.

By definition, **biomass** fuels are fuels that are produced from organic material. These fuels may be used for a variety of purposes, from the heating of homes to fueling the automotive industry. The sources that are used can be classified as either biomass waste or as energy crops. Energy crops are used to convert biomass to such products as bio-diesel, which can be used as a substitute for petroleum diesel fuel to power trucks and heavy equipment. Biomass sources used for heating can include everything from corn to wooden pellets to cherry pits. Biomass as a source of heat should not be confused with bio-fuels, which are fuels such as ethanol that have been derived from a biomass product such as corn. This section will look at the use of biomass as a heating source and discuss how this source can be applied to the heating of residential and commercial structures. In today's world, there is a vast array of organic material that can be used as a viable source of heating fuel.

SOURCES OF BIOMASS FOR HEAT

When choosing an alternative source of energy to heat a home or business, it is important to understand how to effectively apply these types of fuels with the correct combustion equipment and to ensure that the equipment being used is safe and code compliant.

The most popular alternative energy sources of renewable heating fuels include:

- Wood
- Corn

- Wood pellets
- Miscellaneous sources such as cherry pits, rye, and wheat

Wood Characteristics

Using wood as a source of comfort heat can be economical and have a lower impact on the environment as compared to fossil fuels. It can be the sole source of heat for the building, or it can be used as a supplement to the existing primary source. Depending on the type of wood used, it can be the fuel source with the lowest cost per BTU available.

Wood is mostly composed of cellulose, the structural component of cell walls found in green plants, and is also the most common organic compound on earth. About 88% of all wood is composed of cellulose and **lignin**, which is the chemical compound that binds with cellulose to create strong cell walls. When a healthy tree is growing, water constitutes between one-third and two-thirds of its weight. However, when thoroughly air-dried, wood may contain as little as 15% moisture by weight. This is important to remember because the amount of moisture inside the wood will determine whether it will burn effectively and efficiently. In fact, wood that has been dried properly will produce at least 20% more heat output because of its lower moisture content.

Wood is generally classified as either hardwood or softwood. Some examples of hardwood tree varieties are oak, maple, and ash. Pine, spruce, and cedar make up some of the softwoods. This is an important point to understand because the variety of tree will also determine the BTU content of the dried wood. Notice that Table 13-1 lists various types of wood and the heating value of each type in million BTUs per cord. Furthermore, Table 13-2 compares the value of wood in cost per million BTUs to other heating sources.

Table **13-1**

Weight per Cord and BTU per Cord of Air-Dried Wood and the Equivalent Value of No. 2 Fuel Oil in Gallons

Type	Weight Cord	BTU Per Cord Air Dried Wood	Equivalent Value #2 Fuel Oil, Gals.
White Pine	1800#	17,000,000	120
Aspen	1900	17,500,000	125
Spruce	2100	18,000,000	130
Ash	2900	22,500,000	160
Tamarack	2500	24,000,000	170
Soft Maple	2500	24,000,000	170
Yellow Birch	3000	26,000,000	185
Red Oak	3250	27,000,000	195
Hard Maple	3000	29,000,000	200
Hickory	3600	30,500,000	215

Table 13-2

Heating Value of Wood Compared to Other Sources of Heating Fuels

Fuel	Units	Unit BTUs	Unit Cost	Cost Per Million BTUs
# 2 Fuel Oil	Gallons	144,000	$2.60	**$18.06**
Natural Gas	100 cubic ft.	100,000	$1.03	**$10.30**
Propane	Gallons	92,000	$2.15	**$23.37**
Electricity	kWh	3,410	$0.10	**$29.33**
Mixed Hardwood	Full Cord	25,000,000	$200.00	**$8.00**
Live Steam	1,000 Lbs	1,000,000	$28.00	**$28.00**

© CENGAGE LEARNING 2012

Note: Unit costs are based on 2009 figures and may vary according to geographic area.

Things to Know

HOW TO CALCULATE HEATING COST PER MILLION BTUs

The way to make a true comparison of the heating cost for various fuels is to take into account both the heating value of the fuel and the current cost for a unit of that fuel. By using these figures in the following formula, we can calculate the cost for a million BTUs of any given fuel source:

$$\frac{1,000,000 \times \text{Cost of Fuel Unit}}{\text{Heating Value of Fuel Unit in BTUs}} = \text{Cost for 1 Million BTUs of That Fuel}$$

Clearly, wood is a very economical source of heat—even more so if the building owner has a personal source of wood such as a woodlot or forest. A wood stove is the most popular and economical means of burning wood (Figure 13-1). The wood stove can be located almost anywhere as long as there is adequate clearance around the appliance and a chimney can be provided for it. A cord of wood is a stack that measures 4 feet wide by 4 feet high by 8 feet long (Figure 13-2). It is the unit by which wood is bought and sold. A cord of wood can be either split or un-split, and can also consist of mixed varieties. Because wood is sold by volume and not weight, a cord that has been split will consist of more wood because there is a greater amount of surface area (Figure 13-3). Always keep wood covered from the elements if it is stored outside, but allow for air circulation through the stack for proper drying. Wood should be split when it is still green, but only burned when it has properly dried.

One of the biggest issues when using wood for fuel is burning it when it is still green. This practice can result in incomplete combustion, causing the buildup of

Figure 13-1

A wood stove is the most popular means of burning wood for heat.

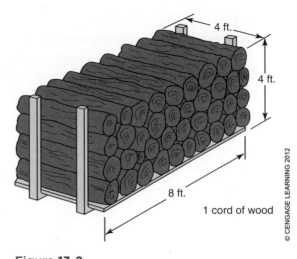

Figure 13-2

This cord of wood is neatly stacked on runners, allowing proper air circulation.

creosote within the chimney and inside the wood-burning equipment. Creosote is a thick, dark liquid when it is hot. When the chimney or flue temperature falls below 250°F, creosote will condense on the surface. When cooled, it forms a tar-like substance that coats the chimney and burner equipment. The buildup of creosote can be very dangerous. It is very flammable and can result in a chimney fire, which can even cause a fire inside the building itself. To prevent the formation of creosote buildup, burn only dry hardwood and maintain an elevated temperature within the chimney. Also, when using an indoor wood-burning appliance, minimize the length of the stove pipe between the burner and the outside wall. The stove pipe should also have a minimum slope of ¼″ per foot of horizontal run (Figure 13-4). This will maintain an elevated temperature in the stove pipe and up into the exhaust flue.

Figure 13-3

Wood that has been split will have a greater amount of surface area per cord.

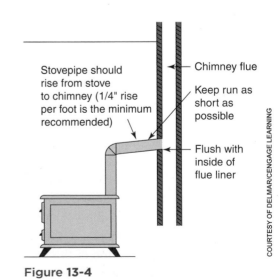

Figure 13-4

Always maintain the proper stovepipe rise between the wood burner and the outside wall.

Characteristics of Corn

Corn is one of the most abundant crops grown in the United States (Figure 13-5). It is also an excellent source of heating fuel. In addition, there are a number of manufacturers of stoves and furnaces available today that specialize in burning shelled corn. Because of these facts, corn can be considered a viable alternative to fossil fuels for comfort heat. A bushel of corn weighs 56 pounds. The average fuel value of corn is approximately 7,000 BTUs per pound. Therefore, a bushel of corn contains a net fuel value of approximately 392,000 BTUs (Figure 13-6).

By inputting an average price of corn at $2.50 per bushel, the cost per million BTUs can be calculated. Using the formula mentioned here, the average cost per million BTUs of corn equals about $6.38. This makes corn even more price competitive than wood! Of course, the cost of heating a given building will fluctuate along with the price of corn. Also, the efficiency of the corn burner will have a significant impact on the overall cost of heating. However, if there is an abundance of corn at a competitive price, it could be considered a very viable alternative compared to other sources of heating.

© ISTOCKPHOTO/PETER GARBET

Figure 13-5

Corn is one of the most abundant crops grown in the United States.

© ISTOCKPHOTO/CYNTHIA BALDAUF

Figure 13-6

The net fuel value of corn is approximately 392,000 BTUs per bushel.

Things to Know

HOW WOOD BURNS

Wood burns in several phases. First, it reaches its ignition temperature, where most of the moisture is driven out. Next, the wood begins to break down chemically when it reaches a temperature of 500°F. Volatile compounds are then released and burned at about 1,100°F. Finally, the remaining charcoal burns slowly at temperatures exceeding 1,100°F.

Emissions are actually higher from smoldering fires than when the fire is raging. This is where we get the term "white-hot coals."

The corn that is to be used as a fuel does not necessarily need to be of the highest quality. However, there are two requirements when using corn as a fuel:

1. The shelled corn must be dry. Preferably, the moisture content should be 15% or less. Corn that is higher in moisture will have a lower heating value per unit weight. Corn that is moist may also have problems with flowing through the fuel-loading auger.
2. The shelled corn must be free of small particulate called *fines*. These are small pieces of cracked kernels and cob pieces that will also cause problems with the fuel-loading auger.

There are several other limitations in using corn as a heating fuel. These include the fact that corn will draw vermin—namely, mice. Furthermore, the corn must be kept in a dry area to prevent the buildup of mold. Therefore, the logistics behind providing a proper storage area for corn must be considered.

Wood Pellets

Wood pellets are another good source of biomass used for heating (Figure 13-7). Most common wood pellets are made from sawdust and ground wood chips (Figure 13-8). These are usually derived from waste materials that are created when making furniture and other wood products. The lignin and other resins found naturally in the cells of wood create the binders that hold wood pellets together, which means no additives are needed. Pellet mills across the United States receive, sort, grind, dry, compress, and bag wood pellets into a conveniently handled fuel. Today, there are over 60 pellet mills throughout North America that produce over 600,000 tons of fuel per year. This figure has more than doubled over the last 5 years. Bagged wood pellets can be purchased at building supply stores, feed and garden supply stores, nurseries, and through wood stove dealers. Pellets may be sold by the bag or by the ton.

Figure 13-7

Wood pellets are a good source of heating fuel.

Figure 13-8

Most common wood pellets are made from sawdust and ground wood chips.

The Pellet Fuels Institute (PFI) located in Arlington, Virginia, sets the standard for pellet quality. PFI represents manufacturers, retailers, and distributors of wood pellet appliances and fuel supplies. Because the chemical and moisture content of other biomass materials may vary, the PFI has developed its own fuel standards for wood pellets. These industry standards assure that the final product maintains as much uniformity as possible for a naturally grown material that is processed but not refined as a fuel. Pellets that are graded by PFI standards must comply with:

- Proper density
- Proper dimension
- Limited amount of "fines"
- Limited salt content
- Limited ash content

The ash content found in wood pellets determines the fuel grade. This is due to the fact that the amount of ash found in pellets will determine the amount of maintenance required to remove the ash from the stove and its venting system. Premium-grade pellets contain less than 1% ash content and are usually produced from hardwood and softwood sawdust that contains no tree bark. These pellets make up over 95% of all current pellet production and can be burned in stoves that call for either a standard or premium fuel.

The heating value of wood pellets can vary from 8,000 to 9,000 BTUs per pound depending on the type of wood that the pellets are made from and the area of the country where they are produced. The price of pellets can range anywhere from $2.50 to $6.00 per 50-lb. bag or from $120 to $200 per ton depending on the region, season of the year, and availability, just like with other heating fuels. At an average price of $150 per ton, the cost per million BTUs of wood pellets is about $11.50, making them competitive with corn or wood as a heating fuel source. The average consumer will use about 40 pounds of wood pellets every 24 hours for each 1,500 square feet of living space being heated. Obviously, this figure will depend on the overall efficiency of the home and its temperature setting.

One of the main appeals of using wood pellets is convenience. Bags of pellets can be stacked compactly and are easily stored (Figure 13-9). One ton of pellets can fit into an area as small as 4 feet wide by 4 feet high by 4 feet long. This is about half the area needed for a cord of wood. They can also be stored in a dry garage, in a basement, or in a utility room or shed. Because of this, consumers will tend to purchase and stockpile pellets in the off-season to take advantage of lower prices, and to make sure they have a sufficient amount before the heating season.

COURTESY OF DONALD STEEBY

Figure 13-9

Wood pellets are typically sold in bags, which makes storing them convenient.

Miscellaneous Sources of Heat

Other sources of combustible materials, such as cherry pits, rye, or wheat, may also be used to heat a home or business (Figure 13-10). The approximate heating value for each or these sources is as follows:

- Dried cherry pits = 9,500 BTUs per pound
- Rye = 7,200 BTUs per pound
- Wheat = 7,160 BTUs per pound

As with other fuel sources, the actual BTU quantity will vary depending on the quality of the product and its moisture content. The practicality of using these materials as heating fuels will vary according to geographic location and product availability. For instance, it may be more advantageous to utilize cherry pits if the building to be heated is located in Washington, California, Oregon, or northern Michigan.

Rye is more predominantly grown in the Midwest region of the United States and in Canada (Figure 13-11). Wheat may be available in most areas of the United States, but only on a seasonal basis (Figure 13-12).

In order to effectively use these products as a heating source, a multi-fuel stove or furnace must be utilized. With this type of appliance, the user is not tied to a particular type of fuel to meet the building's heating needs. This flexibility, combined with a high burning efficiency, will make the multi-fuel appliance quite economical to operate. One of the main advantages of using a multi-fuel stove or furnace is the ability to burn whatever fuel source happens to be readily available

Figure 13-10

Dried cherry pits are also a good source of biomass fuel.

Figure 13-11

Rye can be used as a fuel source.

Field Tip

Mixing Different Types of Fuel Sources

In some instances, it may be advantageous to mix different types of fuel sources to achieve better fuel efficiency and more complete combustion. For instance, because cherry pits burn hotter than other fuels, it is a good idea to mix them with corn or wood pellets. This will create a slower burn and prevent the appliance from becoming overheated.

Figure 13-12
Wheat can be used as a fuel source.

Figure 13-13
A multi-fuel furnace must be used when burning a variety of biomass fuel sources.

at the time that heat is needed. Some multi-fuel appliances can even achieve up to 85% in fuel efficiency (Figure 13-13).

Regardless of whichever type of fuel source is selected, the user will need to match the correct heating appliance to its proper fuel source and ensure that it has been applied and installed correctly.

Chapter 14 | APPLICATIONS FOR BIOMASS BURNERS

When the decision has been made as to which type of biomass fuel to use, the next step is to choose the correct appliance to utilize that fuel most effectively and also learn how to properly install and maintain it. There are numerous different solid-fuel heating appliances available today, and the consumer should be familiar with their capability and how this type of equipment correctly fits into the particular application.

Things to Know

LOCAL WOOD-BURNING REGULATIONS

Certain communities in the United States may restrict the use of wood burning as a means of heating homes and businesses. This is especially true if the air quality in the area is poor. Some of the reasons for these restrictions are because of the smell, the smoke, and the soot, which can be problems for some wood-fired appliances. Although wood burners certified by the Environmental Protection Agency (EPA) may be exempt from these regulations, it is important to first check the local ordinances before considering the use of a wood-burning appliance for heating purposes.

DIFFERENT TYPES OF INSTALLATIONS

The type of installation that the solid-fuel appliance utilizes will depend on a number of factors. For instance, the appliance may be located inside or outside of the home or business. It may be used as a forced-air furnace or a hot water boiler. It may even be utilized with radiant floor heating. Following are the most common applications for biomass heating and their installation practices.

WOOD, CORN, AND PELLET STOVES

Indoor heating stoves are some of the most popular and efficient biomass appliances available. They can be installed almost anywhere as long as there is enough clearance around them and the chimney can be correctly routed to an outside wall.

Buildings with open floor-plan designs and limited separations between rooms are most conducive to indoor heating stoves.

Stoves may be constructed of steel, cast iron, or a combination of these materials. Stoves that are made of steel are generally the least expensive. This type of stove is welded together and contains a firebox made of **refractory bricks**. Steel stoves tend to radiate heat quickly after a fire has been started in them and will give off considerable heat as the fire grows. Cast-iron stoves are more durable than steel stoves and tend to heat more evenly. In addition, cast-iron stoves can be very decorative, offering more intricate detailed artwork incorporated into the frame (Figure 14-1).

Figure 14-1

A wood-burning stove made from cast iron.

© ISTOCKPHOTO/AMY WALTERS

In terms of their heating capability, cast iron stoves hold heat better by cooling down more slowly after the fire dies out. Another material that is used in the construction of stoves is soapstone. This material may be used in combination with cast iron to provide a gentle heat that radiates for a longer period of time than that from other types of materials.

Because stoves transfer heat primarily by radiation, the best installation practice is to locate them centrally in the main floor living area where the occupants spend most of their time. Sizing of the stove is important due to the fact that an oversized stove will operate much of the time with a slow, smoldering flame to keep from overheating the room. On the other hand, a stove that is sized too small can become damaged due to frequent over-firing in order to keep up with the heating demand of the conditioned space.

Ideally, there should be a means of directing the heat to other areas of the structure; however, this may or may not be easy to accomplish. One way to direct heat throughout the structure is to utilize a circulating fan. Notice the wood stove in Figure 14-2. This homeowner installed his wood-burning stove in front of the

Figure 14-2

This homeowner uses a wood stove with a circulating fan above a drop ceiling to heat the house.

COURTESY OF DONALD STEEBY

fireplace in the basement, so he could utilize the chimney. It is also used to heat domestic hot water. Notice that it is near a window where firewood can be easily accessed from outside. The basement has a drop ceiling. Above the wood stove is a circulating fan that forces warm air up from the stove through the space between the drop ceiling and the upstairs floor, keeping the main floor warm. The circulating fan is operated by a thermostat located on the main floor of the house. This application is simple, yet very cost effective, and could be applied to almost any ranch-style home that has an unfinished basement.

With regard to the structural design of the indoor stove, some aspects are related more to aesthetics than performance. There is little difference between the functionally of a cast-iron stove and one that is made from rolled steel, other than that a cast-iron stove may possess more charm. What is different is that some new stoves will shield the sides and top of the unit. All new stoves have shielding on the bottom and rear to prevent overheating and allow closer clearances to combustible walls. Additional shielding provides more heating by convection rather than by radiation, resulting in more heat delivered to the space by warm air rather than by direct radiation from the stove's hot surface. Stoves such as these may need to incorporate a fan to help circulate the warm air throughout the room.

Stove Combustion Design

With the passage of the EPA's wood stove emissions standards in July of 1988, the internal design of indoor heating stoves has changed drastically. The EPA's mandatory emission limit for wood stoves is now only 7.5 grams of smoke per hour. Today, all wood stoves sold in the United States must meet this standard. As a result of this mandate, stove manufacturers have greatly improved the combustion technology of their products, and now some newer stoves have certified emissions of less than 5 grams per hour. The idea behind certifying wood-burning appliances is to achieve more complete combustion. A stove that is not certified tends to starve the fire of oxygen, causing incomplete combustion and creating excessive levels of smoke. In contrast to this, an appliance that is certified will create the right conditions for more complete combustion. These conditions include providing enough combustion air, operating at a higher temperature, and providing sufficient time for the products of combustion to burn more thoroughly before they cool down.

Things to Know

CERTIFIED STOVES

The reason that certified stoves are safer than noncertified ones is because the fire in a certified stove produces less creosote. This is due mostly to a more complete combustion of the wood product. As a result, less creosote means less chance for a chimney fire. These stoves will also tend to reduce the amount of wood that they consume, thus saving the owner money, and the chore of cutting wood.

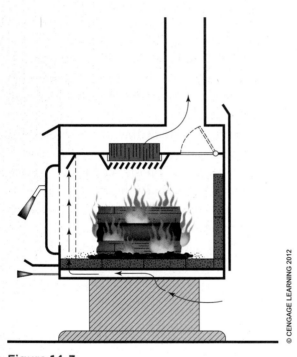

Figure 14-3

A catalytic stove.

© CENGAGE LEARNING 2012

Catalytic and Noncatalytic Stoves

The two main types of certified wood stoves available today are classified as catalytic and noncatalytic. **Catalytic stoves** contain a catalytic converter that acts as an afterburner, increasing the burning of tars, vapors, and other organic compounds that are found in wood smoke (Figure 14-3). The converter is made up of a ceramic honeycomb that is coated with a noble metal—typically platinum, palladium, or rhodium, which are all chemically stable under extreme temperatures. Once the converter is heated to approximately 500°F to 600°F, the smoke from the fire is routed through the converter, where a chemical reaction burns the flue gases. Catalytic stoves are capable of producing long, even heat output; however, the converter will degrade over time and must be replaced every few years.

Noncatalytic stoves force unburned gases through several stages of combustion by forcing the smoke through secondary heat exchangers, where it is mixed with preheated oxygen (Figure 14-4). This process burns the products of combustion at temperatures higher than 1,000°F, which produces more heat and provides a higher operating efficiency. Noncatalytic stoves are more commonly found on the market and are less expensive than catalytic stoves. Even though they are less efficient, noncatalytic stoves are much simpler to maintain. These types of stoves are heavily insulated and contain a large baffle that results in a very hot flow of gas. Many people prefer noncatalytic stoves because they produce a large, roaring fire.

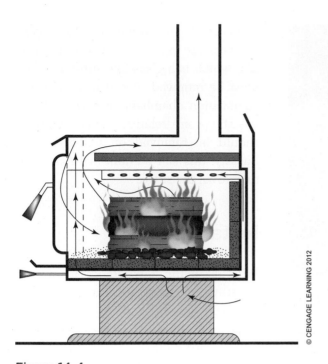

Figure 14-4

A noncatalytic stove.

Corn and Pellet Stoves

Corn and pellet stoves are very similar to wood stoves except that they are specially designed to burn dry granular fuel. These types of stoves are very versatile, and like wood stoves are designed to heat large open spaces.

Unlike wood stoves, corn and pellet stoves are designed to have their fuel supply metered into the burning chamber. This is because both corn and wood pellets are very dense; therefore neither will burn efficiently if they are "piled up" in a combustion chamber. In order to burn effectively and efficiently, the heating fuel is contained in a hopper or storage bin that is usually part of the overall unit (Figure 14-5).

The corn or pellets are then metered into the combustion chamber, where they are burned. The metering process may include an auger to feed the fuel into the combustion chamber, or they may be "dribbled" into the chamber from the hopper above. The feed rate of the fuel is usually adjusted to ensure that the proper heat output is produced. Also, in order to maintain proper combustion, a small fan is incorporated into the system to introduce outside air into the combustion chamber (Figure 14-6).

Most corn and pellet stoves incorporate a circulating fan to help keep the conditioned space warm. As corn or pellets are burned, they produce an incombustible

Figure 14-5

The hopper on a corn-burning stove.

COURTESY OF VERMONT CASTINGS

Figure 14-6

A modern pellet stove.

fragment of ash residue called a **clinker**. Clinkers are made up of silicon dioxide, which is a glass-like substance that must be removed. Because most of the combustion chambers on corn and pellets stoves are relatively small, clinkers should be removed on a daily basis. Luckily, this can be done without shutting down the stove.

Installation

Whether using a wood, corn, or pellet stove, certain procedures must be followed when installing the biomass heating appliance. Failure to correctly install any biomass-fuel-burning appliance can result in unwanted smoke or even a possible fire to the building. Keep in mind that the items listed here are not designed to replace the manufacturer's owner's manual. Rather, they are to provide an overview of safe

Tech Tip
Handling Clinkers

Removing clinkers is a relatively easy task. A specially designed poker device is used to upend the clinker, and then it is removed with a set of tongs.

installation practices. For further information on wood, corn, or pellet stove installation, consult the manufacturer, stove dealer, or local building official.

The three main areas of proper installation practices include:

- The floor and walls around the appliance
- The stovepipe
- The chimney

Clearances

The floor beneath the stove must be covered with a noncombustible base in which to set the stove upon. Typically, a concrete slab with a tile or brick covering will do the trick. Other materials include a prefabricated stove board or mat that is UL approved, or ceramic tile, marble, or slate that covers a UL-listed cement underlayment board. This noncombustible base must be installed under the entire stove and must extend 12″ past each side, and 18″ beyond the front of any loading doors.

In addition to protecting the floor, the stove must also be installed a safe distance from any combustible walls. Depending upon the size of the stove, this distance may vary from 8″ to 36″ or more. Check the owner's manual for the proper distance. As a means of reducing the distance of this clearance requirement, an approved noncombustible wall protection may be installed. This may mean that the adjacent wall is covered with brick, stone, or cement board. An approved UL heat shield may also be added as long as a 1″ air space is provided between the wall and the shield.

The Stovepipe

The **stovepipe** is used to connect the freestanding stove to the chimney, usually by means of a stove collar often supplied by the stove manufacturer (Figure 14-7). The stove pipe is not, however, designed to be installed through walls, floors, or ceilings. Stovepipes are classified as either single-wall or double-wall construction. Single-wall stovepipe is typically made of a heavier-gauge metal than conventional warm air ductwork—usually 22 to 24 gauge. It is also coated with high-temperature-resistant black paint. The rule for clearances with the stovepipe is a minimum of 18″ from any combustible wall, ceiling, or furniture. As with the clearances around the stove, the minimum requirement for the stovepipe clearance can be reduced by using an approved heat shield, or by protecting the combustible surface with an approved method.

A double-wall, or close-clearance, stovepipe is made of a stainless steel inner wall and an outer wall of galvanized metal that is coated with high-temperature-resistant black paint. An air space between the two walls acts as an insulating layer that allows this type of stovepipe to be installed as close as 6″ to combustibles. The double-wall stovepipe is classified for interior use only.

If the length of the stovepipe must be trimmed, use a heavy-duty pair of tin shears.

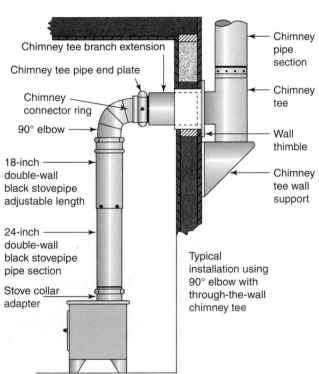

Chimney tee branch extension

Chimney tee pipe end plate

Chimney connector ring

90° elbow

18-inch double-wall black stovepipe adjustable length

24-inch double-wall black stovepipe pipe section

Stove collar adapter

Chimney pipe section

Chimney tee

Wall thimble

Chimney tee wall support

Typical installation using 90° elbow with through-the-wall chimney tee

© CENGAGE LEARNING 2012

Figure 14-7

Detailed drawing of a stovepipe installation.

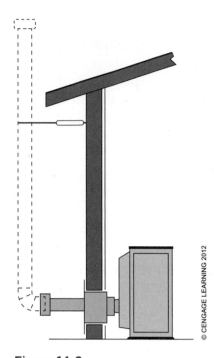

Figure 14-8

A through-the-wall chimney installation.

© CENGAGE LEARNING 2012

When assembling stovepipe, always use three sheet-metal screws at each joint and seal the joint with black furnace cement. Be sure that the pipe is securely fastened to the stove and to the chimney.

The Chimney

The chimney is one of the most important elements to be aware of as far as safety concerns. There are specific rules that must be followed to ensure that the installation is done correctly and safely. The first rule to remember is that the stovepipe cannot be used as a chimney, and it is not allowed to pass through walls, floors, or roofs. For the chimney itself, UL-approved stainless steel, Class A, insulated piping is required. This piping is made of either double- or triple-wall construction and can withstand high temperatures. Several different installation applications can be used with a Class A chimney.

When the stove is located near an outside wall, the venting of exhaust gases may pass through a horizontal vent pipe that includes an end cap to prevent wind from back-drafting into the stove. In this case, the horizontal vent pipe that penetrates the wall must be constructed of type PL vent piping that has been tested to UL 641 standards (Figure 14-8).

This type of vent piping also has double-wall construction. With this configuration, the stove will usually incorporate an exhaust air fan to vent the products of combustion. It is important to ensure that all joints are properly sealed and an end cap is installed when using this configuration so that combustion fumes do not enter the conditioned space should there be a power failure.

Other installation applications for use with Class A chimneys are shown in Figure 14-9.

As mentioned earlier, the stovepipe is not allowed to penetrate through walls or ceilings. When passing through occupied spaces, use a Class A double-wall chimney and maintain the proper clearances to combustibles. Also make sure that the connections between floors and roof are properly supported. When exiting through a wall and continuing up along the side of the building, keep horizontal runs to a minimum and again use the proper chimney material outside of the structure. A minimum rise of ¼″ per foot of run should be incorporated. When exiting the roof, use the "10-2" rule to ensure that the chimney is extended high enough above the peak. This rule states that all chimneys must extend a minimum of 3 feet above the roof surface, 2 feet higher than any part of the roof within 10 feet of the chimney (Figure 14-10).

In summary, things to remember when installing a "Class A" chimney are:

- Do not allow the stovepipe to pass through walls, ceilings, floors, or windows.
- A taller chimney will have better **chimney draft**.
- Keep the number of bends and elbows to a minimum.
- Follow the "10-2" rule.
- Always incorporate an end cap on the top of the chimney to prevent backdrafting and to keep birds and small animals from entering.
- When in doubt, consult the stove dealer or local building official.

Figure 14-9

Three different types of chimney installations.

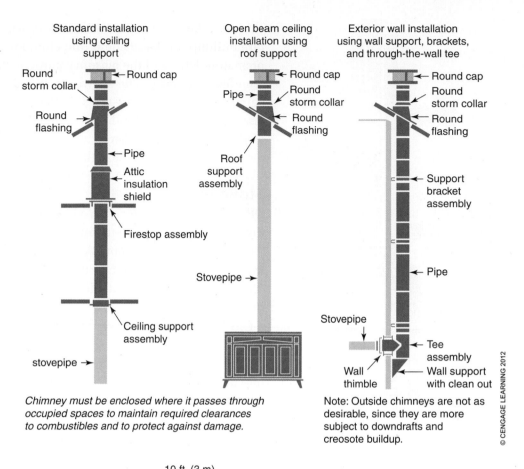

Standard installation using ceiling support

Open beam ceiling installation using roof support

Exterior wall installation using wall support, brackets, and through-the-wall tee

Chimney must be enclosed where it passes through occupied spaces to maintain required clearances to combustibles and to protect against damage.

Note: Outside chimneys are not as desirable, since they are more subject to downdrafts and creosote buildup.

© CENGAGE LEARNING 2012

Figure 14-10

The "10-2" rule for extending a chimney through the roof.

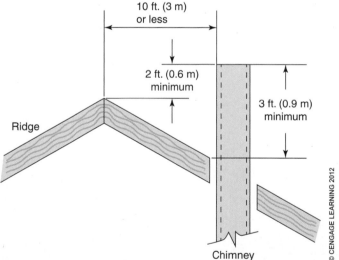

© CENGAGE LEARNING 2012

Masonry Chimneys

Older buildings may already have an existing masonry chimney in place. In some cases, this chimney may be utilized with the biomass stove; however, there are several items that must be considered before doing so. One of the big issues when using an existing masonry chimney is that they are often oversized for use with the stove, which can result in poor drafting and an excessive amount of creosote buildup. This is especially true when using a fireplace chimney. If the products of combustion cannot maintain a high enough temperature when passing through the masonry

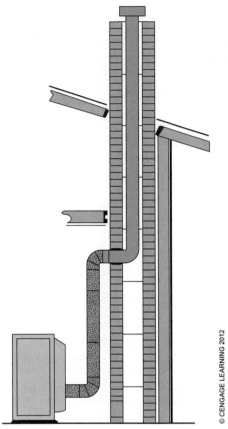

Figure 14-11

A chimney liner should be installed when utilizing an existing masonry chimney.

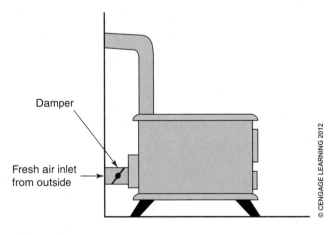

Figure 14-12

A stove using a make-up air tube from outside.

chimney, they will begin to condense, which will create moisture buildup on the inside of the chimney. This can result in a premature failure of the masonry joints, causing the chimney to crumble and decay. The answer to this dilemma is to install a chimney liner (Figure 14-11). This flexible liner is made of stainless steel and resembles the vent pipe that is used on residential clothes dryers. Remember to consult the stove manufacturer for the recommended liner size that matches the stove.

Stove Safety

Other issues to consider with a biomass stove are the incorporation of a source of make-up air and the use of smoke detectors. Air that is used for combustion in the stove must be supplied from outside, either directly or indirectly. For instance, an additional inlet from an outside wall can be attached directly to the appliance, as shown in Figure 14-12.

Older buildings may have leaks through doors, walls, and windows that may supply enough combustion air indirectly without requiring additional make-up air sources. However, modern buildings that are well sealed and insulated may require provisions for additional make-up air. Remember that not providing enough air for combustion can result in health hazards to the occupants.

Smoke detectors are a must for any home or business, and in most cases are required by local fire codes. They should be incorporated into all buildings regardless of the type of heating appliance being used. It is recommended that all smoke detectors be hard wired into the building's 120-volt circuitry and include a battery backup.

Stove Maintenance

Like any other type of heating appliance, biomass stoves require some maintenance. These appliances usually require more maintenance than conventional gas-fired appliances due to the residue and ash that is produced from burning a biomass product. The unit should be inspected on an annual basis to ensure that all components are functioning properly and safely. This includes the inspection of such things as the venting, fans and motors, and feed auger. Inspect the stovepipe and chimney for creosote buildup before and at least once during the heating season. Further routine maintenance includes cleaning out the inside of the stove on a weekly or even on a daily basis, depending upon the stove's usage. Use a wire brush to remove the buildup of ash and soot in and around the combustion chamber. Remove ash deposits from the ash drawer and vacuum out the burner chamber to help maintain the efficiency of the appliance. Be sure to use a quality fuel product to help ensure more complete combustion. Be sure to keep the phone number of a qualified service technician handy in case problems arise that cannot be handled by the homeowner or building owner.

OUTDOOR BOILERS

Another method of heating a home or business with biomass is to install an outdoor boiler. This appliance can use wood or corn as a fuel source, although wood is the predominant choice. The advantages of using an outdoor heating appliance are rather obvious. They include:

- Reduces the risk of a fire to the premises
- No dirt, soot, or insect problems because the fuel is stored outdoors
- No risk of indoor air pollution from a fire burning inside the building
- Typically there is no increase in property insurance by having the unit outdoors

Installation of Outdoor Boilers

Outdoor boilers are often installed between 50 and 100 feet from the building (Figure 14-13). This is to reduce the risk of fire to the structure and, more important, to improve the venting of the products of combustion. A boiler that is located too close

Figure 14-13

An outdoor wood boiler, loaded and ready to fire up.

COURTESY OF DONALD STEEBY

Figure 14-14

A water-to-air heat exchanger.

© CENGAGE LEARNING 2012

to the structure can allow smoke and soot to enter the structure, causing problems with indoor air quality. With an outdoor boiler, water is used as a means of heat transfer. The boiler heats water that is transferred through underground piping to the indoors and through a water-to-air heat exchanger (Figure 14-14).

This heat exchanger is mounted in the supply plenum of the forced-air furnace (Figure 14-15).

Unlike a typical residential hot water boiler, it is important to understand that this configuration is not a pressurized system. The water that is circulated through the outdoor boiler is vented to the atmosphere. This is due to safety reasons—specifically, to prevent an explosion in the event that the outdoor boiler runs out of water. A pressurized hot water system requires additional safety features and falls under a separate set of building code regulations. Another important point to emphasize is that an outdoor boiler typically circulates a water and glycol mix to guard against freezing.

COURTESY OF DONALD STEEBY

Figure 14-15

A heating coil mounted in the supply plenum of a residential furnace.

Sizing the Boiler

The best method of determining the proper sized boiler is to perform a heat loss calculation on the structure. A detailed description of performing this task can be found in Chapter 11 under the heading "Load Calculations." Once this calculation is known, consult the boiler manufacturer or dealer to find out the efficiency of the chosen model. Traditionally, outdoor wood- or corn-fired boilers maintained a heating efficiency of about 50%. This factor has improved significantly over the past several years as the demand for this type of heating has increased. If a water-to-air heat exchanger is used inside of the structure, this component will need to be sized as well. It is best to consult an HVAC wholesaler or heating coil manufacturer to assist in sizing the heat exchanger. These sources can determine the proper dimensions of the coil to fit the existing plenum and ensure that the proper heating output is achieved.

Setting the Boiler

Outdoor boilers can weigh as much as 3,000 pounds when filled with water. Therefore it is important to ensure the boiler is mounted on a solid base. Typically, a 4″ concrete pad is poured to support the boiler. Most manufacturers will provide a template with their product that will show the required dimensions of the concrete pad (Figure 14-16).

The amount of concrete may vary, but typically about ½ yard will suffice. Remember that the pad should extend beyond the front of the boiler to allow the

Figure 14-16

The foundation plans for an outdoor wood boiler.

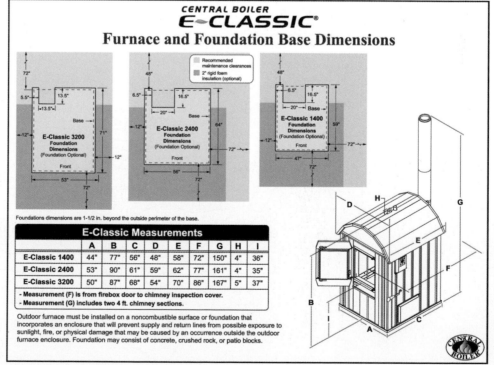

operator ample space for loading the firebox. An alternative to pouring concrete for a base is the use of concrete blocks. An ample number of blocks should be used at each corner of the boiler, or they should be set around the complete perimeter of the boiler. Be sure that these blocks rest on a firm foundation in the ground to prevent shifting of the boiler once the blocks settle.

Things to Know

CHOOSING THE RIGHT SIZE BOILER

Be careful not to choose an outdoor wood boiler that is oversized. Problems will occur during mild weather when there is a minimal demand for heat, and the wood simply smolders inside the combustion chamber without really burning. Also, choose a boiler with a large access door. This will make it easier for loading wood into the boiler.

Piping Installation

The first step in running the underground piping between the boiler and building is to dig a trench that is below the frost line. This will prevent excessive heat loss and prevent water freezing in the pipes if the system does not use antifreeze. The frost line is the maximum depth that frost will penetrate the ground during the coldest period of the heating season. This depth will vary depending upon geographic location. Figure 14-17 shows typical frost line depths throughout the United States.

Figure 14-17

U.S. map showing frost line depths.

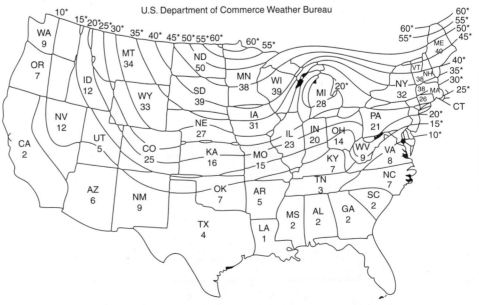

COURTESY OF U.S. DEPARTMENT OF COMMERCE, WEATHER BUREAU

The actual digging will require the use of a trencher, ditch-witch, or small backhoe that can be rented by the do-it-yourselfer. Otherwise, contract with a local excavator, landscaper, or septic tank installer.

The pipe of choice for underground installation is called PEX. This stands for cross-linked polyethylene. PEX is flexible and can withstand high- and low-temperature fluctuations. PEX is also easy to install and is highly resistant to chemicals found in the plumbing environment. The smooth interior of PEX will not corrode and is also very freeze and break resistant. Although it is not mandatory to install PEX lines below the frost line, doing so will prevent a large amount of heat loss through the system. Two lines of 1″ to 1¼″ PEX will need to be installed for supply and return water. These lines should be encased in high-density urethane insulation and wrapped in a high-density polyethylene jacket to prevent any underground moisture or frost from coming in contact with the hot water lines (Figure 14-18).

If this situation should happen, substantial heat loss will result, which will significantly affect the amount of hot water that will be available to heat the building. Another suggestion is to run the hot water piping through oversized PVC pipe. Remember that the PEX piping will expand and contract when heated and cooled, and the PVC pipe will allow for this without affecting the pipe's performance. At the very least, use an oversized sleeve through the foundation wall where the piping enters the building to allow for expansion and contraction.

Field Tip

Installing a Boiler in Winter Months

An outdoor boiler can be installed during the winter when the ground is frozen. Simply lay the insulated water lines on the ground on top of a bed of straw and cover the lines with more straw to help insulate them. There will be some heat loss, but it should be minimal. Then, when spring finally comes, the water lines can be buried.

Figure 14-18

The underground piping used with outdoor boilers. Pipe is marked to easily identify supply and return lines.

Thermo PEX with 1″ central PEX lines

Thermo PEX with 1–1¼″ central PEX lines

Connect the PEX piping to the indoor heat exchanger using special compression fittings provided by the piping supplier. These fittings will require a special tool that can also be provided by the PEX supplier. At the outdoor boiler, the piping will be connected to a circulating pump. On larger outdoor boilers, or if the boiler is a significant distance from the building, it is not unusual to use two circulating pumps, one on both the supply and return lines.

Wiring and Controls

When installing the piping in the outdoor trench, include the electrical wiring to run the boiler. This wiring is typically a 12-gauge, 3-conductor, 110-volt cable that should be run in its own ¾″ PVC electrical conduit underground. Another suggestion is to install a nylon cord through both the electrical and hot water PVC sleeves in case an additional circuit is needed in the future. Also install a new 15-amp circuit breaker into the building's breaker panel and connect the boiler wiring to both the indoor breaker and at the outdoor unit. The boiler's outdoor wiring should include a lighting fixture and convenience outlet. This circuit will also power the circulation pumps.

The controls for the outdoor boiler are relatively simple. They are divided between the outdoor and indoor controls. Most boilers are controlled outdoors by a line-voltage thermostat that senses the hot water temperature (Figure 14-19). When this temperature falls below its setpoint, the contacts on the thermostat close, which in turn opens a damper on the outdoor unit. This damper allows combustion air to enter the fire chamber and increase the flame, thus raising the

Figure 14-19

Thermostat controls for an outdoor boiler.

COURTESY OF DONALD STEEBY

water temperature. When the thermostat is satisfied, it closes the damper and flame is reduced to a smolder. In any event, there is always a fire burning in the outdoor unit.

Indoors, a room thermostat senses the space temperature and, on a call for heat, energizes the furnace circulating fan. The fan blows air across the heated hot water coil, warming the space up to its temperature setpoint. When the space temperature is satisfied, the fan is de-energized.

The last control scenario deals with the circulating pump. There are two schools of thought with regard to controlling the pump. One says the pump should be energized continuously. This is true if there is no antifreeze solution in the water; otherwise the piping would freeze. If the water does contain antifreeze, the circulating pump could be cycled on and off either by the outdoor thermostat, which controls the water temperature, or whenever the indoor space thermostat calls for heat. It would be best to cycle the pump with the hot water thermostat; otherwise there may not be hot water present in the indoor coil when there is a call for heat by the indoor thermostat. The pump could also be controlled by a thermostat that is strapped to the supply line going into the indoor coil to ensure this coil always has hot water (Figure 14-20).

Figure 14-20

A strap-on thermostat used to control the circulating pump.

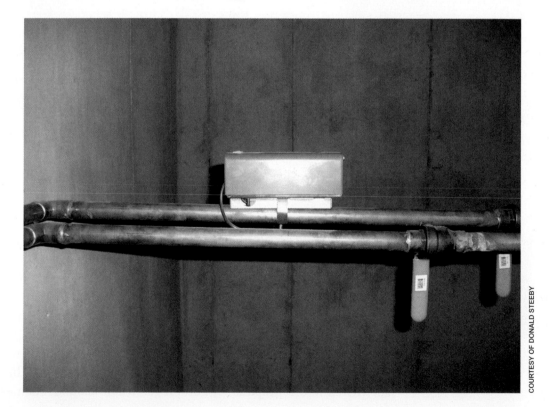

COURTESY OF DONALD STEEBY

Additional Applications

If the outdoor boiler is using corn or wood pellets as a fuel source, a storage hopper will be required. This hopper may be integral to the boiler or externally mounted (Figure 14-21).

Green Tip

Using an Outdoor Air Reset Control

Some modern boilers come equipped with a device used to reset the hot water temperature setpoint from the outdoor air temperature. These dual temperature controllers sense both the outdoor air temperature and the hot water supply temperature. Here is how they work: As the outdoor air temperature increases, the boiler's hot water temperature setpoint automatically decreases. Lowering the hot water setpoint during milder weather saves money and reduces wear on the boiler. There is still some debate as to whether to use this type of temperature controller on an outdoor boiler. Some say it is not worth the additional cost because a wood-fired boiler is more efficient when it is operating at a higher temperature.

Figure 14-21

Storage bins used with outdoor boilers for corn and wood pellets.

It should be of significant size so that it does not need to be filled too often. In addition, the corn boiler will require an auger device to feed corn into the combustion chamber. Some modern corn and pellet boilers include electronic controls that automatically ignite the fuel source at each heating cycle. These controls also can incorporate self-modulating adjustments for low, medium, and high fire settings depending on the output heating demand. Modern outdoor boilers can also include dual-fuel capability. In addition to biomass, these units can switch to oil, propane, or natural gas when needed, such as in an emergency situation.

Indoors, the boiler can be utilized for other heating applications as well. These include:

- Heating of domestic hot water by using a heat exchanger (Figure 14-22)
- The use of radiant floor heating or baseboard heat instead of forced air
- Pool and hot tub applications

COURTESY OF DONALD STEEBY

Figure 14-22

A heat exchanger used for heating domestic hot water.

- Greenhouse heating applications
- Heating multiple buildings

Boiler Maintenance

The maintenance on an outdoor boiler is similar to that of the indoor stove. This maintenance includes such items as inspection of the combustion chamber for wear and inspection of the chimney for creosote buildup. Also ensure that controls are in good working order and are properly calibrated. An overall seasonal cleaning is always recommended to ensure the equipment is performing to its fullest potential.

In addition to these items, the water supply must be maintained to ensure proper system efficiency. The system's water level should be checked periodically to ensure it is at its proper level. Add water to the system when necessary. To do this, a permanent water line should be connected from the building's domestic water supply to the water lines of the boiler as a means of adding make-up water. The water lines may also need to be intermittently purged of any air that becomes trapped in the lines. If the system includes water filters, these must be cleaned periodically as well.

One more maintenance task is to remove ash from the outdoor unit as needed, usually once or twice per month during the heating season. Some units include an ash auger that automatically empties the outdoor unit of ash buildup.

INDOOR BOILERS AND FURNACES

Indoor appliances used to burn biomass fuels follow the same sizing, installation, and maintenance requirements as do stoves. The difference is that these appliances can be used as either the primary source of heat for the building or may be added on to the existing source.

When used as a primary source of heating, the indoor biomass furnace will typically be a dual-fuel or multi-fuel appliance. In addition to using wood, corn, or wood pellets as a biomass source, these furnaces may also use oil, propane, natural gas, or even coal as an alternate fuel source. It would be prudent to install a multi-fuel unit when it is to be used as a primary source of heat so there is always a backup source available in emergencies.

Multi-fuel Furnaces for Primary Use

Multi-fuel or combination furnaces have separate combustion chambers that are side by side yet still connected (Figure 14-23). The first stage of the furnace ignites the wood portion, which burns until the space is warm, at which time the burner cycles off. This sequence occurs until the wood supply is depleted. If no more wood is added, the furnace will operate as a conventional gas or oil furnace until more wood is added to the firebox. This way the building is never without heat. These types of furnaces incorporate a two-stage thermostat that controls both combustion chambers at separate temperature settings. If the wood portion of the furnace dies down, the second thermostat setting will ignite the fossil fuel, automatically maintaining the building at the temperature set for that fuel source.

Figure 14-23

An example of a multi-fuel furnace.

COURTESY OF YUKON-EAGLE FURNACE COMPANY

Add-on Furnaces

Add-on biomass furnaces are also used in conjunction with the fossil-fuel source. In this scenario, the biomass furnace is installed as a retrofit to the existing furnace. These furnaces may also burn wood, corn, or wood pellets as a source of fuel. Typical installation involves connecting the supply and return plenums of the add-on furnace to the building's existing ductwork. The furnace is hand-stoked if it is a wood-burning appliance, or the hopper is filled with corn or wood pellets if it uses those types of fuels (Figure 14-24).

The controls for add-on biomass furnaces include the use of two separate space thermostats. Typically, the primary thermostat controls the biomass furnace, and is set to maintain the desired room temperature. A second thermostat controls the fossil-fuel furnace that acts as a backup. If the fire goes out on the biomass unit, the room temperature will drop until it reaches the lower setpoint of the

Figure 14-24

An example of an add-on corn-burning furnace.

COURTESY OF DONALD STEEBY

© CENGAGE LEARNING 2012

Figure 14-25

A draft gauge is used to check for proper venting.

secondary thermostat. The fossil-fuel furnace then takes over to maintain the space temperature setpoint of the secondary thermostat.

Another control strategy for use with add-on furnaces includes the use of isolation dampers located in the supply and return ductwork. These are motorized dampers that close when the fossil-fuel furnace is in operation to prevent heated air from circulating through the biomass furnace.

Venting requirements for the biomass furnace follow the same rules as for biomass stoves. Maintenance requirements are also the same. One separate maintenance requirement for biomass furnaces involves the use of a **draft gauge**. This device is used to test the proper rate of combustion air in barometric pressure over the fire and through the chimney (Figure 14-25).

The furnace should always maintain a negative differential pressure over the fire and through the chimney in relationship to the building pressure. Typical pressure readings should be −0.02 inches of water column pressure over the fire, and −0.04 to −0.06 inches of water column pressure through the chimney. Adjustments to the amount of primary air entering the furnace may need to be made to achieve these settings. Always consult the manufacturer's recommendations when making any adjustments.

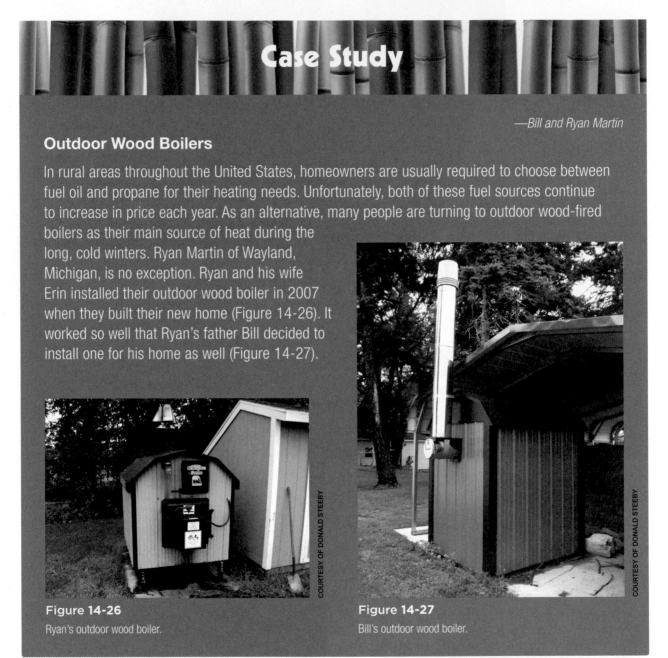

Case Study

—Bill and Ryan Martin

Outdoor Wood Boilers

In rural areas throughout the United States, homeowners are usually required to choose between fuel oil and propane for their heating needs. Unfortunately, both of these fuel sources continue to increase in price each year. As an alternative, many people are turning to outdoor wood-fired boilers as their main source of heat during the long, cold winters. Ryan Martin of Wayland, Michigan, is no exception. Ryan and his wife Erin installed their outdoor wood boiler in 2007 when they built their new home (Figure 14-26). It worked so well that Ryan's father Bill decided to install one for his home as well (Figure 14-27).

Figure 14-26
Ryan's outdoor wood boiler.

COURTESY OF DONALD STEEBY

Figure 14-27
Bill's outdoor wood boiler.

COURTESY OF DONALD STEEBY

Ryan's boiler heats his single-story home (which is approximately 5,000 square feet) with about 12 cords of wood per heating season. Bill, on the other hand, heats about 12,000 square feet (which includes his farm house and a guest house) with about 20 cords of wood each winter. Both men do not mind cutting wood each year, and they have a bountiful supply of trees on the acres that they own.

Ryan did most of the installation of his wood boiler himself, saving him an immense amount of money. He figures that his total installation cost was about

Figure 14-28

Manifold piping for the radiant floor heating system.

Figure 14-29

Large access door on boiler.

Figure 14-30

Boiler control panel with light-emitting diode (LED) display.

$10,000. When he factors in the cost of propane over the last 3 years, he estimates that his system paid for itself in a little over 2 years—an outstanding payback period. The Martins get the maximum value from their system by utilizing the wood boiler for not only comfort heat, but also for heating their domestic hot water. Their boiler even provides radiant floor heating in their finished basement as well (Figure 14-28).

Ryan does point out a couple of things that he prefers on his father's outdoor boiler. For one, Bill's boiler has a larger access door, making it easier to load wood (Figure 14-29). Another item is the advanced control panel that even has a temperature controller to allow for the adjustment of the operating hot water temperature (Figure 14-30).

Both families have acquired quite a collection of chain saws and wood-cutting equipment. They even make their annual wood-cutting event a family affair. With the money that they are saving in heating costs, cutting and stacking wood every year is almost a pleasure.

UNIT 5

Future Energy Sources: A Look at Fuel Cells and Combined Heat and Power (CHP)

Chapter

15

HOW FUEL CELLS WORK

INTRODUCTION

Up to this point, the most current sources of alternative energy have been examined. The wind, the sun, and the earth are the most viable sources of alternative energy, and they all have been successfully used in both residential and commercial applications. There are, however, other sources of alternative energy that need to be discussed. These sources may not receive the same exposure as some of the more popular choices that are available today, but they do possess the same characteristics. The most redeeming characteristics of alternative energy sources include:

- They are plentiful.
- They are renewable.
- They are clean.
- They don't harm the environment.

Two sources of alternative energy that are gaining popularity are **fuel cells** and combined heat and power (CHP). Let us first examine the fuel cell.

The history of fuel cells dates back to 1839, when Sir William Grove began experimenting with the electrolysis of water. He discovered that hydrogen and oxygen could be combined to produce water and an electric current. From 1889 until the early 20th century, many people tried producing fuel cells that could convert coal directly to electricity. Unfortunately, these attempts failed because the experimenters did not know enough about the materials necessary to create fuel cells. It was not until 1932 that Francis Bacon and his research team developed the first successful fuel cell. This design was eventually used by the NASA *Apollo* moon missions and also by the Skylab programs (Figure 15-1).

Today, fuel cell technology is being used to power everything from cars to homes and businesses. This unit will focus on the use of stationary fuel cells to generate energy for residential and commercial usage.

WHAT IS A FUEL CELL?

A fuel cell is a device that converts a chemical reaction into electricity. In most cases, this chemical reaction is the combining of hydrogen and oxygen together to form electricity. Rather than combusting a fossil fuel to generate electricity, fuel cells rely on the process of reverse electrolysis to develop the same result. The difference between electrolysis and reverse electrolysis is that during the process of electrolysis, an electric current is applied to water, which produces hydrogen and oxygen (Figure 15-2). By reversing this process, separate sources of hydrogen and

Figure 15-1

An alkaline fuel cell
for a shuttle orbiter.

Figure 15-2

The electrolysis process.

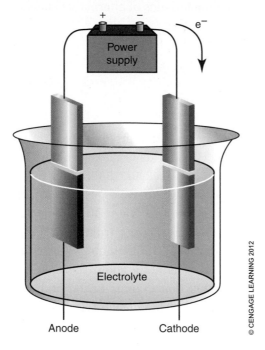

oxygen are combined together to produce electricity and water as a by-product. Other by-products of fuel cells include heat and carbon dioxide.

Fuel Cell Composition

Fuel cells are made up of three layers. One layer consists of an **anode**, which is defined as a terminal where current flows inward. Another layer consists of a **cathode**, where current flows out. Between these two layers is a polymer membrane electrolyte. This membrane contains a catalyst that facilitates the reaction of the hydrogen and oxygen within the fuel cell.

The anode is considered the negative side of the fuel cell, and it performs several jobs. For one, it conducts electrons that are released from the hydrogen molecules so they may be used in the electrical circuit. The anode typically has channels etched into it to allow the equal dispersion of hydrogen gas over the surface of the catalyst (Figure 15-3).

The cathode is considered the positive side of the fuel cell. It too has channels etched into its surface that distribute the oxygen molecules to the surface of the catalyst. It also conducts electrons back from the electrical circuit to the catalyst, where they recombine with the hydrogen and oxygen to form water.

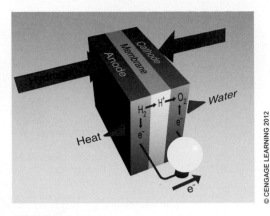

Figure 15-3

The parts of a fuel cell.

© CENGAGE LEARNING 2012

The electrolyte within the membrane is where the proton exchange takes place. This electrolyte conducts only positively charged ions, but blocks the electrons. The catalyst is a special material usually made of platinum. It is rough and porous so that there is a maximum amount of surface area to be exposed to the hydrogen and oxygen molecules.

How a Fuel Cell Works

Fuel cells produce electrical power in the form of direct current without the need for combustion. They are similar to a battery in that they consist of two electrode plates, an anode and a cathode. However, with a fuel cell, these plates are separated by a polymer membrane electrolyte. Also unlike a battery, fuel cells are constantly replenishing their power output.

Here is how the fuel cell creates an electrical charge: Hydrogen gas is introduced into the anode, which is made of platinum. The platinum creates a catalytic reaction that ionizes the hydrogen gas. This ionization splits the hydrogen atom into positive and negative ions. Both of these types of ions are naturally drawn to the cathode, which is positioned on the opposite side of the membrane. However, only the protons are allowed to pass through the membrane. Therefore the electrons are forced to go around the membrane, where they create an electrical circuit. While the hydrogen gas is being introduced to the anode, oxygen from the air is being fed to the cathode. Here the catalyst is creating oxygen ions. When the hydrogen protons and electrons bond with these oxygen ions, the resulting waste product is water (Figure 15-4).

A single fuel cell produces only about 0.7 volts of electricity. In order to increase this voltage, manufacturers stack fuel cells together in series. More layers equal higher voltage output. However, in order to increase the current output, the fuel cell must have a larger surface area.

Figure 15-4

How a fuel cell works.

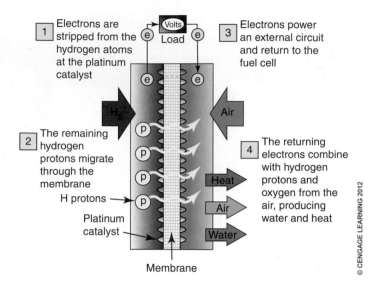

1 Electrons are stripped from the hydrogen atoms at the platinum catalyst

2 The remaining hydrogen protons migrate through the membrane

3 Electrons power an external circuit and return to the fuel cell

4 The returning electrons combine with hydrogen protons and oxygen from the air, producing water and heat

H protons

Platinum catalyst

Membrane

© CENGAGE LEARNING 2012

Advantages of Fuel Cells

As long as the hydrogen is being supplied, the fuel cell will continue to generate power. Because this power is produced by an electrochemical process, there are no products of combustion as compared to burning fossil fuels. The fuel cell process is clean, quiet, and highly efficient. In fact, fuel cells are two to three times more efficient than conventional fossil fuels. When the waste heat from fuel cells is utilized along with their power-generating capability, they become **cogeneration** units. In larger building systems, these fuel cell cogeneration systems can reduce energy costs by 20% to 40% over conventional fuel energy systems, while increasing efficiency as much as 85%. Fuels cells that are used for stationary power cogeneration achieve a fuel-to-electricity efficiency of up to 40%. When pure hydrogen is used as a fuel source, a fuel cell becomes a zero-emission power source. In fact, some fuel cell power plants are so low in emissions that some areas of the United States have exempted them from air permit requirements. Other advantages include:

- Fewer moving parts, which results in less maintenance
- Reliability: No power loss during an electrical storm
- Quiet operation means a reduction in noise pollution
- Less dependence on foreign oil

Applications

There are more than 2,500 fuel cell systems being used throughout the world. Their applications include hospitals, nursing homes, hotels, office buildings, and schools. Fuel cell systems are being used as both a supplement to the existing

Figure 15-5

A commercial fuel cell.

© CENGAGE LEARNING 2012

power grid and also as stand-alone systems in areas that are inaccessible to power lines (Figure 15-5).

Fuel cells for residential and light commercial applications can also be used as a primary source of power or as a reliable backup to grid-supplied electricity. Just as with solar photovoltaic (PV) and wind-turbine-powered generators, fuel cells can be controlled independently or in parallel with the existing power grid. They can be installed in the basement of a building or in the backyard. A fuel cell for an average home or business is about the size of a domestic refrigerator. Newer fuel cells can even extract hydrogen from a variety of conventional fossil-fuel sources such as propane and natural gas. Some utility companies will offer to lease fuel cells for domestic usage, or the building owner may opt to purchase a unit outright. Just as with solar PV and wind turbine systems, excess power that is produced by the fuel cell can even be sold back to the electric utility company.

One issue in dealing with stationary fuel cells for residential and commercial use is how to deliver pure hydrogen to the place where the fuel cell is located. Even though hydrogen is the most abundant element in the universe, it is quite volatile and can often be difficult to extract from other compounds. Fortunately, the transportation and distribution of pure hydrogen gas is nothing new. Hydrogen

Things to Know

PROPANE FOR RESIDENTIAL FUEL CELLS

Many fuel cell manufacturers are considering the use of propane as a hydrogen-carrying fuel source for rural and remote fuel cell applications. The target market for this application consists of widely dispersed homes and small businesses that already use propane and that are located far from main power lines and gas mains.

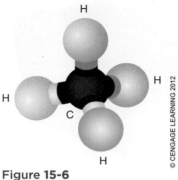

Figure 15-6

A methane molecule consisting of four hydrogen atoms and one carbon atom.

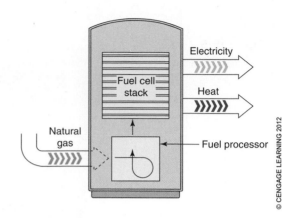

Figure 15-7

A fuel cell with a built-in fuel processor for natural gas.

has been transported for industrial use since the early part of the 20th century. Storage methods for transporting hydrogen typically consist of steel cylinders that hold the hydrogen in a gaseous state and can withstand pressures of up to 2,000 pounds per square inch, in much the same way that propane is distributed. However, where does the homeowner or business owner obtain pure hydrogen on a regular basis, and how is it to be stored at the fuel cell's location? One solution to this dilemma is to obtain the hydrogen needed to power the fuel cell from an alternative source, such as from a hydrocarbon. Although fuel cells use pure hydrogen as a primary fuel source, many companies are developing fuel cells that use natural gas to power their product. This conversion from natural gas to hydrogen is performed by the use of a fuel processor. One of the reasons why natural gas is the choice for powering fuel cells is because many customers already have natural gas delivered to their homes and businesses. Another reason is because natural gas is primarily made up of methane, which has the chemical analysis of CH_4 (one carbon atom and four hydrogen atoms; Figure 15-6). Because of its composition, it is easier for the fuel cell converter to "strip out" the one carbon atom from natural gas and then be left with four atoms of pure hydrogen.

The fuel processor is what makes the fuel cell practical for residential and commercial applications. Its task is to take the natural gas or propane and convert it to hydrogen that is pure enough to be utilized by the fuel cell—typically with no greater than 50 parts per million of carbon monoxide. Simultaneously, the fuel processor must meet emission levels for the other by-products of the conversion process (Figure 15-7).

FUEL CELL INSTALLATION

Though most fuel cells for residential and light commercial use can fit comfortably inside a mechanical room or outside a home or business, they are typically not installed by the "do-it-yourselfer." Most fuel cell manufacturers will require that their product be installed strictly by factory-trained and authorized personnel.

A typical 2,000-square-foot home will require a 5- to 7-kilowatt fuel cell to meet its electrical needs. This type of fuel cell will be about the size of a chest freezer (Figure 15-8).

Figure 15-8

A residential fuel cell application.

© CENGAGE LEARNING 2012

Because fuel cells operate at approximately 150°F, they can also use their waste heat to produce hot water or auxiliary space heat for the home, thereby increasing the overall efficiency of the system. Fuel cells for the home or business can also be connected to the utility grid through the main service panel, much like a solar photovoltaic system or wind-powered generator. The local utility provider may also offer a net metering service that can assist in recovering the cost of the equipment. When combined with the existing electrical service, fuel cells can provide continuous backup power by means of an automatic switchover in the event of a grid interruption (Figure 15-9).

This backup power can be used to power-up dedicated circuits in the home or business while the grid is down, then automatically switch back when power is restored. One important point to remember is that fuel cells produce direct-current (DC) power. This direct current must be converted to alternating-current (AC) power

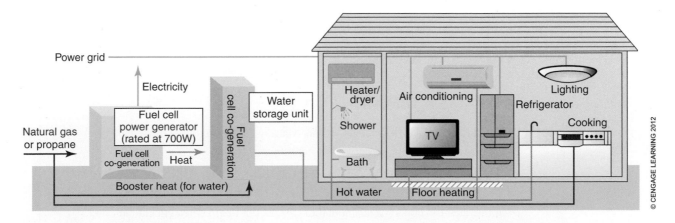

Figure 15-9

A fuel cell can be used for residential power and heat.

Figure 15-10

Simplified residential
fuel cell system.

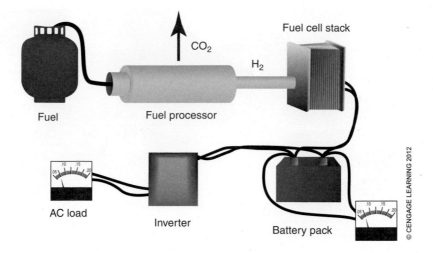

before it can be utilized by the home or business, just like with solar and wind power. Therefore, the implementation of a charge controller and inverter must be utilized with the system. Furthermore, a battery pack may need to be incorporated to provide backup power when the fuel cell is not operational (Figure 15-10).

Costs

When comparing a dollar-per-watt installation cost, most residential and light commercial fuel cells are comparable to a solar photovoltaic system. However, fuel cells of similar size and application can generate up to eight times more energy per year than the same size solar PV installation, even in the sunniest locations. As a comparison, a 5-kW solar photovoltaic system will generate approximately 10 megawatts of power per year. However, the same 5-kW fuel cell can produce up to 80 megawatts of combined electrical and waste heat energy. Another difference between these two systems is that the fuel cell will require natural gas as an operating cost, whereas the solar panel requires no fuel to operate. The typical operating cost for a residential or commercial fuel cell can be as low as 6% per kilowatt-hour based on a natural gas price of $1.20 per **therm**. One way to appraise the value of a fuel cell system is to

Things to Know

WHAT IS A THERM?

A therm is equal to 100,000 BTUs. It is typically used to compare the pricing of natural gas. The energy content in 1 cubic foot of natural gas is approximately 1,050 BTUs. However, this number may vary depending upon the quality of the gas. In order to equalize the pricing based on BTU content, most gas utilities will use either cost per hundred cubic feet (CCF) or cost per therm when pricing their fuel.

Figure 15-11

Homeowners and business owners can receive state and federal incentives to install fuel cells.

compare its costs and benefits against competing energy technologies. Although the cost of fuel cells continues to decline, they still have some way to go before they are affordable for the average business or household.

Fuel cells used for homes and businesses, however, can be eligible for substantial economic rebates as part of state and federal renewable energy policies. Current federal tax incentives can cover up to 30% of the cost to purchase and install qualifying fuel cells for residential usage (Figure 15-11). These incentives do not expire until the end of 2016. In addition, states such as California will offer rebates of up to $2,500 per kW of installed power through the Self-Generation Incentive Program (SGIP) through the end of 2011. Businesses may further benefit from the installation of fuel cells by taking advantage of accelerated depreciation programs offered by the federal government. The Database of State Incentives for Renewables and Efficiency (DSIRE) website provides comprehensive information on state and federal incentives for the promotion of renewable energy systems.

FUEL CELL MAINTENANCE

Typically, fuel cells require less maintenance than most sources of alternative energy (Figure 15-12). In fact, some manufacturers recommend one planned maintenance service call each year. This, however, would be for a backup system only. Because fuel cells have no real moving parts, maintenance is focused on resupplying the hydrogen fuel. If the fuel cell has a fuel processor that extracts the hydrogen from natural gas or propane, this device will receive the most attention. Most of the diagnostics within the fuel cell depend on the level of sophisticated controls it contains. Some may have automatic monitoring systems that tell the owner or technician if and when there is a potential problem. Other control features include run-time monitoring, self-testing diagnostics, and unattended "conditioning cycles" to ensure that they are operational when needed. If the

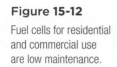

Figure 15-12

Fuel cells for residential and commercial use are low maintenance.

fuel cell is being used strictly as a power backup system, especially if used for a mission-critical application such as a hospital, codes and safety standards may require that periodic inspections be more frequent, such as on a monthly basis.

LOOKING TO THE FUTURE WITH FUEL CELLS

Following are some of the issues that the fuel cell industry will need to resolve before fuel cells can become the power-generating appliances for homes and businesses that will replace conventional energy-generating systems:

- **Hydrogen supply**: It is important to remember that fuel cells are not like batteries. Although they both share the ability to change chemical energy into electrical energy, fuel cells require an adequate supply of pure hydrogen in order to operate properly. Although there is an abundance of hydrogen found in nature, it must be extracted from other molecules. Free hydrogen is volatile and can be difficult to store. If there is not an abundance of fossil fuel available from which to extract the hydrogen, there will need to be a local source from where it can be purchased and properly transported to the fuel cell's location.

- **Fuel cell stack humidity**: The stack of fuel cells that are connected together to create a complete unit convert hydrogen and oxygen into electrical power. These two elements must be delivered to the polymer membrane in a continuous and uniform manner. This process can be disrupted if the humidity level in the fuel cell is too high or too low. Excessive humidity can cause the membrane to become clogged, which will inhibit the movement of protons. If the humidity becomes too low, the membrane dries out, rendering the fuel cell inoperable.

- **Cost**: Just as with other forms of energy generation, the market is driven by supply and demand. Once manufacturers can introduce fuel cells to the energy market that are price competitive with conventional sources of heat and power, the average homeowner and business owner can justify implementing them into their building's power network. Just as it is with other forms of alternative energy, owners need to substantiate the additional cost of where they purchase their kilowatts and BTUs that are used to light and heat their buildings.

Chapter | # WHAT IS CHP?

16

INTRODUCTION

Combined heat and power (CHP) is also known as **cogeneration**. This process consists of the simultaneous production of electricity and heat from a single fuel source. Modern CHP facilities can achieve fuel efficiencies of up to 90%. In comparison, conventional power plants that only produce electricity are about 35% to 55% efficient at best. One cause of these low efficiencies is that the waste heat produced by the plant is not easily transportable; therefore it is not utilized. The source of fuel for CHP facilities can be natural gas, propane, coal, biomass, or fuel oil. Whichever kind of fuel is used, CHP can offer a clean, efficient, and reliable source of alternative energy for residential and commercial applications as well as for industry (Figure 16-1).

History of CHP

The first modern use of combined heat and power was by Thomas Edison at his 1882 Pearl Street Station power plant. This first commercial power plant generated both electricity and thermal energy by using waste heat to warm the neighboring buildings. This recycling of waste heat allowed Edison's power plant to achieve approximately 50% efficiency.

By the beginning of the 20th century, steam was the main source of mechanical power. Many of the power houses that produced steam soon realized that they could produce electricity as well. However, as centralized power plants emerged—managed by regional utilities—so did federal regulations to promote

Figure 16-1

A basic configuration showing how cogeneration works.

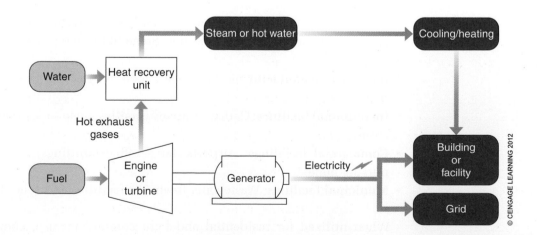

© CENGAGE LEARNING 2012

rural electrification. These regulations discouraged decentralized power generation, such as cogeneration. As a result, larger utility companies became a more reliable and inexpensive source of electricity, so the smaller power houses stopped their cogeneration practices and bought electricity from the larger utilities.

Because of the energy crisis of the 1970s, interest in cogeneration began to revive, and by the end of the 1980s, the need to conserve energy became a necessity. The United States finally passed legislation to encourage the development of cogeneration facilities in 1978. The Public Utility Regulatory Policies Act (PURPA) encouraged cogeneration technology by allowing power plant facilities to connect with the main utility grid to buy and sell electricity. Under PURPA, cogeneration facilities would be allowed to purchase electricity from the main utility companies at a fair price when needed, while also allowing them to sell their electricity based upon the utility's cost to produce the power. This act encouraged utilities to purchase power from other energy producers. As a result of this legislation, a rapid increase in cogeneration capacity developed throughout the United States.

Things to Know

COGENERATION IN EUROPE

In Europe, cogeneration is not given the same government support that it has been given in the United States. This is because Europeans do not recognize it as a new technology. However, surprisingly, cogeneration is used much more frequently in Europe than it is in the United States, especially for residential and light commercial usage.

CHP APPLICATIONS

The technology behind combined heat and power systems can be utilized for a vast array of energy-intensive facilities. Some of these types of facilities include:

- **Industrial manufacturing**: Chemical, food processing, pulp and paper, and petroleum.
- **Institutional facilities**: Colleges and universities, hospitals, prisons, and military bases.
- **Commercial buildings**: Airports, large office buildings, hotels, casinos, and nursing homes.
- **Municipal facilities**: Wastewater treatment facilities and schools.

When utilized for residential and light commercial use, combined heat and power can result in an integrated energy system rather than a single technology.

Figure 16-2

A CHP unit sized for a typical residential application.

COURTESY OF MARATHON ENGINE SYSTEMS

This energy system can also be modified depending on the needs of the individual user. A typical CHP unit for residential or commercial application will range in capacity from about 1 kW up to 6 kW and have the capability of producing around 12,000 BTUs of heat at 1-kW output simultaneously. A 6-kW unit can provide up to 10 gpm of hot water at 140°F to 150°F. This amount of heat could satisfy the heating requirements of a 2,500-square-foot home (Figure 16-2).

The main components of a combined heat and power system typically are (Figure 16-3):

- A fuel-driven internal combustion engine and electrical generator
- A warm air furnace, or hot water boiler
- A heat exchanger that circulates fluid from the generator's engine coolant system
- A control module that operates the system

1	Control box
2	Flue gas heat exchanger with integrated catalytic converter
3	Engine unit
4	Generator
5	Silencer
6	Heat exchanger heating system
7	Electronic connections
8	Gas supply
9	Fresh air/flue gas
10	Heating connections

COURTESY OF DELMAR/CENGAGE LEARNING

Figure 16-3

A cut-away view of a CHP showing its various components.

The operation for a CHP system is as follows:

1. A reciprocating engine drives a generator, producing electricity.
2. This engine is powered by burning fuel—natural gas, propane, or biogas.
3. This electricity is used by the building to offset the cost of purchasing power from the local utility company.
4. A heat recovery system is utilized to capture waste heat from the combustion system's exhaust, or from the engine's coolant system.
5. This heat is converted into useful thermal energy by means of a heat exchanger—either to heat the air through a furnace or heat the water used in a boiler.

Other applications for CHP include the heating of domestic water, or utilizing the waste heat for air conditioning (Figure 16-4). The same configurations for hot water generation that are used with solar thermal storage can be applied to CHP systems. Furthermore, refer to Chapter 3 regarding the application of absorption cooling. The same principles for solar cooling can be applied to CHP systems as well.

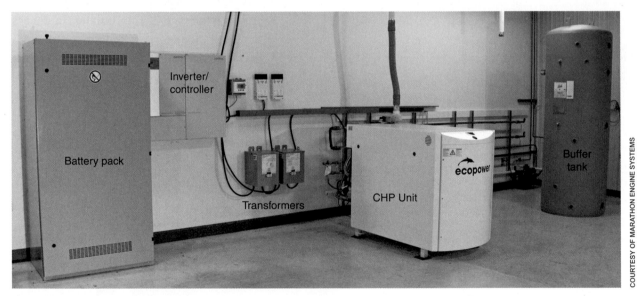

Figure 16-4

Various configurations for CHP systems.

CHP Controls

An important point to recognize is that the technology of CHP for residential and commercial usage is designed to displace the building's existing heating system. The heat that is generated to warm the building is actually a by-product of the electrical generator's cooling or exhaust system. This is how the CHP system is able to maximize its efficiency.

Most residential and commercial CHP systems utilize an auxiliary source of heat within the building, such as a furnace or boiler. These auxiliary sources are

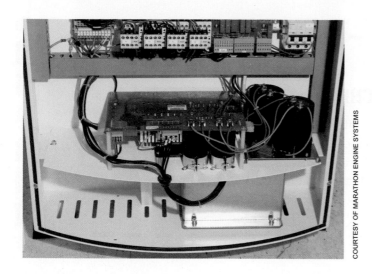

Figure 16-5

Microprocessor controls for a CHP system.

Figure 16-6

An example of an electronically commutated motor (ECM).

automatically operated by the CHP system's control module so that the building stays warm, even on the coldest days of the year, when the CHP system cannot keep up. The CHP system typically incorporates a microprocessor controller that communicates digitally with other components within the system (Figure 16-5).

This controller initially operates the generator portion of the CHP system, delivering electrical power to the structure on a continuous basis. The controller also interfaces with a programmable thermostat to maintain the selected space temperature setpoint. When there is a call for heat, the thermostat activates either the circulating fan located in the warm air furnace or a circulating pump located on the hot water boiler. This same thermostat can operate the cooling mode of the system by activating an absorption cooling unit on demand. If cooling is not used, a means of getting rid of excess waste heat could be an issue. One solution for utilizing excess heat is to use it in conjunction with domestic hot water heating. Other sources of heat sinks include swimming pools and spas.

As an additional energy-saving feature, some furnace units incorporate an **electronically commutated motor (ECM)** on the circulating fan for space heating. This type of motor uses a variable-speed drive that operates at about 10% of the energy consumption as compared to conventional fan motors (Figure 16-6).

What is unique about some CHP systems is that they deliver a smaller amount of space heating on a continuous basis as compared to conventional heating systems that cycle on and off. Delivering a lower-temperature heat output to the space on a continuous basis results in two additional benefits:

1. Electrical savings are realized because the fan motor is not constantly cycling off and on.
2. The conditioned space feels more comfortable due to the fact that the temperature of the air delivered from the heating source is closer to that of the room temperature. This reduces the amount of heat stratification and lowers the amount of temperature swing.

Tech Tip
Internet Connection for CHP

Some CHP manufacturers are now offering connection of their control system directly to the Internet. This allows the customer to remotely monitor the climate control of the home or business. Some parameters, such as time scheduling, can even be changed via an Internet connection from anywhere in the world (Figure 16-7).

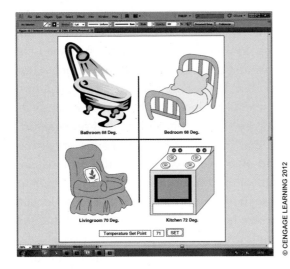

Figure 16-7

An example of an Internet connection to a residence.

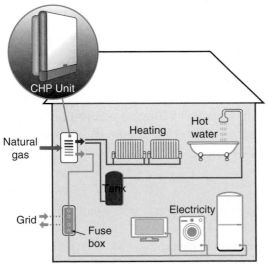

Figure 16-8

The best use for CHP systems is when the waste heat usage can be maximized.

DESIGN GUIDELINES

The following items should be considered when designing a CHP system for a residential or commercial application.

First, the most prudent application for CHP is when the structure can utilize the waste heat on a year-round basis. Without this factor, the system is just generating electricity and the full benefits of CHP are not realized. Under these conditions, the efficiencies could be reduced as much as 50% versus a system that can utilize the waste heat all of the time. A good application that could overcome this situation is a grocery store using an absorption chiller to maintain cold storage. In this application, the heat from the CHP could help satisfy the cooling load during warm months. As mentioned earlier, always look for other sources to utilize waste heat, such as heating domestic hot water as well as swimming pools and spas (Figure 16-8).

Another consideration is to size the system to meet the heating needs. This will optimize the system so that all of the waste heat is being utilized. This will correspond to a 100% utilization factor. The utilization factor is the ratio of waste heat that is being used compared to the total reclaimable amount of heat that is generated. One suggestion is to analyze the heat load on a monthly basis and choose a CHP unit that has a BTU output equal to the lowest month. By doing this, the utilization factor will always be 100%, and any excess heat will be used by conventional means.

When sizing the CHP system, try to achieve a thermal efficiency of at least 60%. In this case, the thermal efficiency is defined as the ratio of the recovered heat plus the amount of electricity generated divided by the fuel input. Anything below a 60% thermal efficiency will not be economically feasible. On the other hand, more efficient systems will satisfy a larger heating load but will produce less electricity.

As a general rule, the CHP should be in full operation for a minimum of 5,000 hours per year in order to assure its financial feasibility and payback.

INSTALLATION PRACTICES

Many new CHP units have been designed with the heating and air conditioning trades in mind. Some units require no additional installation skills beyond those that are already utilized by electricians, pipe fitters, and HVAC installation technicians. Consult the manufacturer or distributor of the CHP unit to become familiar with its proper installation procedure before beginning the work. As always, be sure that the proper mechanical and electrical permits have been obtained before beginning installation. In addition, always adhere to local, state, and national codes for installation of this type of appliance. Once the proper unit has been chosen, several considerations need to be made before installing a CHP system for home or business use. These considerations include:

- Choosing an installation site
- Making the proper electrical connection
- Performing the waste heat interconnection
- Connecting to a backup heating system

Choosing an Installation Site

A residential or commercial CHP is about the size of an average domestic refrigerator. As mentioned earlier, they may be installed outside of the building, in the basement, or in a mechanical room. Wherever the site selected, make sure it is accessible to fuel connections and electrical connections, and is installed with the proper clearances around it. If installing the unit outdoors, be sure it is properly weatherproofed. Because most CHP units operate at a very quiet level, the installation location should not affect any neighboring occupants. Be sure that the unit is mounted on a solid surface and is not subjected to any shifting of its foundation.

Making the Proper Electrical Connection

Just as with solar photovoltaic and wind energy systems, it is important to ensure that the proper electrical connections are installed correctly. When considering the wiring configuration, the building owner must first decide if the system is to be a stand-alone or grid-connected system.

If the system is to be stand-alone, a battery storage package must be included. This would also include the implementation of a charge controller and inverter to convert the DC electrical power to AC power (Figure 16-9). If the unit is to be grid-tied, an inverter would still need to be included (Figure 16-10).

With either configuration, the use of an inverter is essential to ensure that the voltage produced from the CHP unit matches the proper voltage and frequency of the building's electrical system. The manufacturer's instructions should outline how to perform the proper electrical connections at the unit. All electrical connections to the inverter, battery storage package, main service panel, and the building's breaker panel should be performed by a qualified licensed electrician. Review Chapter 5 on solar photovoltaic system wiring and applications for more details on various electrical configurations and their installation.

Figure 16-9

A typical CHP connection using a battery backup system.

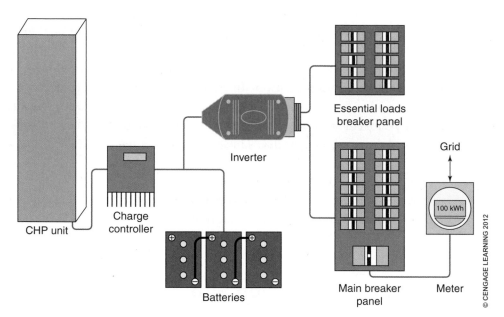

© CENGAGE LEARNING 2012

Figure 16-10

A CHP using a grid-tied connection.

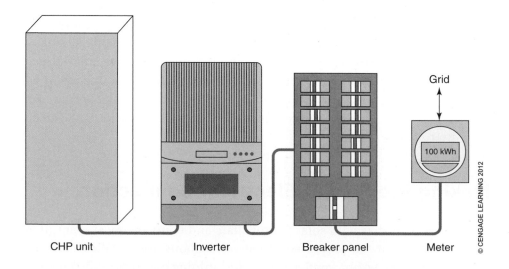

© CENGAGE LEARNING 2012

Performing the Waste Heat Interconnection

This portion of the installation process is generally done by a plumber or pipe fitter. Interconnection of the waste heat from the CHP to the indoor unit is performed in much the same way as for solar thermal storage systems. The piping configuration will depend on whether a warm air furnace or hot water boiler is used as the building's main source of heat.

If using a warm air furnace, the piping will be done from the CHP to a heat exchanger located in the furnace (Figure 16-11).

Typically, a circulating pump will be installed to pump the transfer fluid from the CHP to the furnace. Because the CHP is in operation most of the time, this pump will also be energized to remove the waste heat away from the CHP. When there is a call for heat, the circulating fan on the furnace will be energized, allowing warm air to be circulated throughout the conditioned space.

Piping will be installed between the outdoor unit and indoor heat exchanger in much the same way as if the building is utilizing a hot water boiler for comfort heat (Figure 16-12). Again, the circulating pump between the indoor and outdoor coils will be energized almost continuously while the CHP is in operation. One exception will be the use of a secondary pump that circulates water from the indoor boiler to the heating terminal units. This pump will be energized by the room thermostat whenever there is a call for heat (Figure 16-13).

Figure 16-11

Connecting the CHP heating loop to a residential furnace.

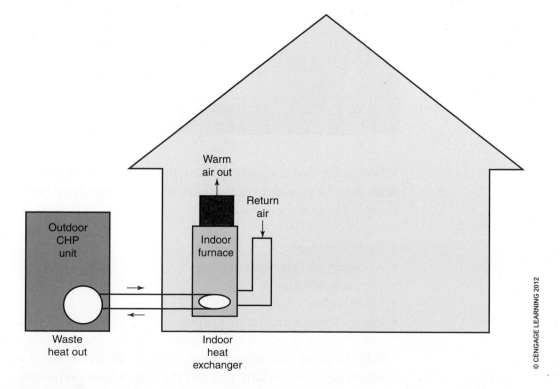

© CENGAGE LEARNING 2012

Figure 16-12

Connecting the CHP heating loop to a residential boiler.

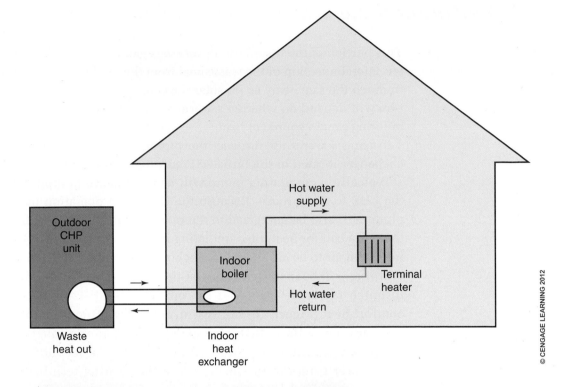

Figure 16-13

A CHP unit installed with residential hot water heating.

The piping that connects the CHP to the indoor heat exchanger will include the same devices as with a conventional heating system, or as used with solar thermal storage. These devices include:

- Expansion tank
- Air vents and air separator
- Pressure-relief valve
- Check valve
- Pressure and temperature gauges

Keep in mind that an antifreeze solution should be used for the heat-transfer medium between the CHP coil and the indoor heat exchanger.

Connecting to a Backup Heating System

As mentioned earlier, a backup heating system is almost a necessity when incorporating a CHP system into the building's central heating system. Even if the CHP is sized for satisfying the building's design heating load, there may be instances when backup heat is necessary, especially if the CHP is in need of maintenance or repair. One advantage of installing a backup system is that it will only be operating for a fraction of the time versus if it were the main source of heating, thus saving the building owner money. Another advantage is that it typically will not need to be sized for the maximum building load. In most instances, the backup heating system incorporates the heat exchanger that is used with the CHP unit into one package. Also, the logic controls used with the CHP system will operate both the main source of heat and the auxiliary backup system automatically (Figure 16-14).

Figure 16-14

A residential CHP installation with hot water backup.

COURTESY OF ECR INTERNATIONAL

Things to Know

RETROFITTING CHP UNITS

Many residential and commercial CHP systems are designed for not only new construction, but also to replace existing heating systems (Figure 16-15). The proper BTU output of the new unit will depend on the size of the existing unit and its level of efficiency. Be sure to take into consideration the level of thermal insulation in the existing building, and always perform a comprehensive heat load study before selecting the new unit.

Figure 16-15

These CHP units may be used with new construction or as retrofits for existing heating systems.

COURTESY OF MARATHON ENGINE SYSTEMS

SYSTEM COSTS

The initial cost of a complete CHP system can include the following items:

- The generator unit
- Engineering and design fees
- Installation costs
- Controls package
- Electrical equipment, including switch gear and disconnects
- Plumbing components, including valves, buffer tank, and fittings

Some manufacturers estimate the cost of installing their unit at approximately twice as much as that of a conventional heating system. The unit cost for a 2-kW to 6-kW system will run between $10,000 and $20,000, with the additional installation cost of approximately $4,000 for new homes. These costs will be comparable to installing a new geothermal system.

Operational costs and energy savings will vary according to geographic area and by the cost of the fuel used to run the system. Additional variables for estimating operating costs include the amount of electricity that the system produces and whether net metering is available in the area where the system is to be installed. As an example, a residential structure located in the Northeast area of the United States with a 1.2-kW system can produce approximately one-half of its annual electricity needs. Taking into consideration the cost of operating the CHP unit at its full capacity, the savings will equal about one-half of the cost of fuel gas and electricity required to produce the same amount of electricity. This is the case as long as the cost of electricity remains above $0.85 per kilowatt-hour. If this same

area has net metering available, the electrical costs will be reduced even further. Remember that to achieve full potential savings, the building owner must be utilizing the full amount of thermal output in the form of waste heat from the CHP unit.

Payback on Investment

The estimated payback for a CHP system will vary according to fuel cost, electricity cost, waste heat utilization, and the availability of net metering. Because heat is in greater demand in colder areas of the country, the best and fastest payback period seems to be in colder climates with higher electric rates and reasonably priced natural gas rates. However, the underlying variable lies in the ability to utilize the waste heat effectively. Systems that can capitalize on waste heat usage, such as for use in swimming pools and absorption cooling systems, will help enhance the payback period. One manufacturer suggests that if the end user is paying more than 14¢ to 16¢ per kilowatt-hour, the CHP unit will pay for itself in a reasonable amount of time. Use the Energy Savings Calculator located at http://www.marathonengine.com as an aid in determining the cost recovery for a given system (under the Cogeneration tab, click on Ecopower and then Savings Calculator).

As a further incentive, building owners should access the federal and state websites for green energy legislation to find out if tax rebates are available for the CHP system they are considering. Federal tax incentives can be researched at http://www.energy.gov (type in "Tax Breaks" in the search bar). State incentives can be accessed at http://www.dsireusa.org.

MAINTENANCE OF CHP SYSTEMS

Most of the maintenance requirements for a CHP system revolve around the gas-fired engine that runs the electrical generator. The extent of maintenance depends on the type of engine and

Green Tip

What Is Net Metering?

As explained in Chapter 5, net metering is a means for the building owner to receive full value for the excess power that is produced by the CHP system. By the use of net metering, a home or business can offset the cost of its electric bill with any extra electricity that is produced. Here is how it works: When the CHP system produces electricity, the power is first used to meet any electrical needs that the building requires. When more electricity is produced than what is needed, the excess power is fed back onto the main utility grid. This causes the electric utility meter to literally spin backward, allowing the customer to receive credit for the additional energy. At the end of the billing period, the utility company credits the customer with the net amount of kilowatt-hours produced at the wholesale power rate. If the customer uses more electricity than the CHP system generates, the customer pays the difference back to the utility company.

COURTESY OF MARATHON ENGINE SYSTEMS

Figure 16-16

This CHP engine needs only regular maintenance once per year.

fuel that is used. Most natural-gas-fired internal combustion engines require routine maintenance every 4,000 hours. This equates to servicing the engine once per year (Figure 16-16).

At the 4,000-hour interval, engine maintenance consists of oil and filter change, spark plug replacement, and other minor adjustments. This type of service call generally takes about an hour and costs about $200. Most manufacturers have authorized dealers who will perform these services under contract.

Most natural-gas-fired internal combustion engines run cleaner than conventional gasoline engines and typically have an engine life of up to 40,000 hours or 10 years.

Other routine maintenance on the CHP system will depend on the peripheral devices that are included. Electrical devices should be checked for any wear and tear, and should have all connections routinely checked for tightness. Plumbing components should be checked for any sign of leaking. Antifreeze should be checked annually to ensure it is valid. Indoor components should be given the same scheduled maintenance as any typical residential or commercial heating and air conditioning equipment.

When properly serviced and maintained, combined heat and power systems should provide the homeowner or business owner with years of economical, trouble-free power.

GLOSSARY

A

absorption cooling a process in which cooling is accomplished by the evaporation of a volatile fluid, which is then absorbed in a strong solution, then desorbed under pressure by a heat source, and then re-condensed at a temperature high enough that the heat of condensation can be rejected to an exterior space.

active systems active solar hot water systems employ a pump to circulate water or heat-transfer fluid between the collector and the storage tank.

air separator a hydronic system component that separates air from water as it flows through the system.

air vent a fitting used to vent air manually or automatically from a hydronic system.

alternating current (AC) an electric current that reverses its direction at regular intervals.

alternative energy energy sources, such as solar, wind, or nuclear energy, that can replace or supplement traditional fossil-fuel sources such as coal, oil, and natural gas.

ambient (temperature) the temperature of the surrounding environment; technically, the temperature of the air surrounding a heating or cooling medium.

american wire gauge (AWG) the standardized wire gauge system used in the United States and Canada for the diameters of round, solid, nonferrous, electrically conducting wire.

ammeter a meter used to measure current flow in an electrical circuit.

ampacity the maximum amount of current a cable can carry before sustaining immediate or progressive deterioration.

amp-hours a unit of electric charge that is transferred by a steady current of 1 ampere for 1 hour.

anemometer an instrument used to measure the velocity of air.

anode a terminal or connection point on a semiconductor.

asynchronous generator a three-phase, cage-wound generator, also called an induction generator, used to generate alternating current. When used on a wind turbine, this type of generator is characterized by the slow, almost constant rate of the blades turning and can be fed directly into the electrical grid without the need of an inverter.

B

bimetal strip two dissimilar metal strips fastened back to back.

biomass organic matter, especially plant matter, that can be converted to fuel and is therefore regarded as a potential energy source.

brazing high-temperature (above 800°F) melting of filler metal for the joining of two metals.

C

capillary tube a fixed-bore metering device used in refrigeration and air conditioning systems.

catalytic stove a stove that contains a cell-like structure consisting of a substrate and catalyst that produces a chemical reaction, causing pollutants to be burned at much lower temperatures.

cathode a terminal or connection point on a semiconductor.

charge controller a device that limits the rate at which electric current is added to or drawn from electric batteries. It prevents overcharging and may prevent over-voltage, which can reduce battery performance or lifespan and may pose a safety risk.

check valve a mechanical device that normally allows fluid (liquid or gas) to flow through it in only one direction.

chimney draft   air that is naturally entrained to flow upward through the chimney by convection buoyancy, which allows products of combustion to be vented from the appliance.

circuit breakers   a device for interrupting an electric circuit to prevent excessive current, as that caused by a short circuit, from damaging the apparatus in the circuit or from causing a fire.

clinkers   the incombustible residue, fused into an irregular lump, that remains after the combustion of biomass such as corn or wood pellets.

closed-loop system   used with solar thermal storage, a closed-loop system is a sealed piping system normally filled with glycol or antifreeze that is sealed off to the atmosphere.

cogeneration   simultaneously generating both electricity and useful heat by means of a power station or heat engine, such as a turbine generator. Also known as combined heat and power (CHP).

combined heat and power (CHP)   the use of a heat engine or a power station to simultaneously generate both electricity and useful heat. Also known as cogeneration.

compressor   a vapor pump that pumps vapor (refrigerant or air) from one pressure level to a higher-pressure level.

condenser   the component in a refrigeration system that transfers heat from the system by condensing refrigerant.

conduction   the transfer of heat between two parts of a stationary system, caused by a temperature difference between the parts.

convection   the transfer of heat by the circulation or movement of the heated parts of a liquid or gas.

creosote   an oily liquid that coats the lining of chimneys and is found in the smoke of wood that has been burned.

cross-linked polyethylene pipe   see PEX.

cubic feet per minute (CFM)   used to determine the rate of flow through a forced-air system.

D

darrieus wind turbine   a type of vertical axis wind turbine (VAWT) used to generate electricity from the energy carried in the wind.

direct current (DC)   electricity in which all electrons flow continuously in the same direction.

diversion controller   controllers that are used to divert excess energy production, typically from solar PV panels to prevents the panel from overloading.

doping   a method of adding an element to a pure semiconductor to change its electrical properties.

draft gauge   a gauge used to measure very small pressures to compare them with the atmosphere's pressure. Used to determine the flow of flue gas through a chimney or vent.

drainback system   a closed-loop solar heating system in which the heat-transfer fluid in the collector loop drains into a tank or reservoir whenever the booster pump stops to protect the collector loop from freezing.

dry well   a well used for the discharged water in an open-loop geothermal heat pump.

E

electrical substation   a subsidiary station of an electricity generation, transmission, and distribution system where voltage is transformed from high to low, or the reverse, using transformers.

electrolyte   a conducting medium in which the flow of current is accompanied by the movement of matter in the form of ions.

electromagnetic   a coil of wire wrapped around a soft iron core that creates a magnet.

electromotive force   a term often used for voltage that indicates the difference of potential in two charges.

electronically commutated motor (ECM)   a DC motor that uses electronics to commutate the rotor instead of brushes. Typically built for motors less than 1 HP.

equipment grounding   the practice of connecting the non-current carrying metal portion of the equipment to a grounding wire to earth.

evacuated-tube collector   a type of solar collector that consists of parallel rows of glass tubes connected to a header pipe. Each tube has the air removed from it to eliminate heat loss through convection and radiation.

evaporator   the component in a refrigeration system that absorbs heat into the system and evaporates the liquid refrigerant.

expansion tank   a small tank used in closed water heating systems and domestic hot water systems to absorb excess water pressure, which can be caused by thermal expansion as water is heated, or by water hammer.

expansion valve   the component between the high-pressure liquid line and the evaporator that feeds the liquid refrigerant into the evaporator.

F

flagging   a phenomenon where tree limbs and branches only grow with the prevailing winds.

flat-plate collector   a solar collector that consists of an insulated metal box with a glass or plastic cover and a dark-colored absorber plate. Solar radiation is absorbed by the absorber plate and transferred to a fluid that circulates through the collector in tubes. In an air-based collector, the circulating fluid is air, whereas in a liquid-based collector it is usually water.

freeze-protection valves   devices that are installed on solar hot water systems providing a thermally controlled freeze-protection system. They contain precise thermal actuators that open the valves when the temperature approaches freezing and close them upon warm-up. This continuous process of modulation, cold water being discharged and replaced by warmer water, prevents panel and piping freeze-up.

frequency   the cycles per second of the electrical current supplied by the power company. This is normally 60 Hz in the United States.

fuel cell   an electrochemical system in which the chemical energy of a fuel is converted directly into electrical energy.

fuse   a protective device, used in an electric circuit, containing a conductor that melts under heat produced by an excess current, thereby opening the circuit.

G

geoexchange   a geoexchange heating and cooling system uses the consistent temperature of the earth to provide heating, cooling, and hot water for both residential and commercial buildings.

geothermal   a type of heat pump that uses the ground, ground water, or ponds as a heat source and heat sink, rather than outside air.

gin pole   a rigid pole with a pulley on the end used for the purpose of lifting. The lower portion of the pole is attached to the upper exterior of an existing tower or structure. The free end extends above the existing tower or structure.

glycol   an antifreeze solution used in the water loop of geothermal heat pumps or in the piping of solar thermal storage systems.

glycolic acid   the by-product of propylene glycol when subjected to extreme heat as a result of chemical transformation.

governor   a device for maintaining uniform speed regardless of changes of load, as by regulating the supply of fuel or working fluid.

grounding   a conducting connection between an electric circuit or equipment and the earth or some other conducting body.

H

heat gain/loss calculation   the sum total of BTUs transferred into and out of a building as a result of convective, conductive, and radiant heat transfer.

horizontal axis wind turbine (HAWT)   a wind turbine in which the axis of the rotor's rotation is parallel to the wind stream and the ground.

horizontal ground loop   a type of closed-loop geothermal heat pump system in which fluid-filled plastic heat exchanger pipes are laid out in a plane parallel to the ground surface.

hydronic collector   a device used to collect, absorb, and transfer solar energy to a working fluid, such as water or air.

I

infiltration loss   heat loss through the building due to openings around doors and windows.

inline fan   a ventilation fan that is placed directly in the system ductwork used to transfer air.

insolation   the amount of solar energy, direct or diffuse, reaching a surface per unit of time. More precisely, insolation is the solar power density incident on a surface of stated area and orientation, usually expressed as watts per square meter (W/m²) or BTU per square foot per hour.

integral collector storage unit (ICS)   a solar thermal collector in which incident solar radiation is absorbed by the storage medium.

internal heat gain   internal heat gain comes from the collection of heat given off by sources inside of a building. The most common sources of internal heat gain are appliances, electronic devices, people, and lighting.

inverter   a device that converts direct-current electricity (for example, from a solar module or array) to alternating current (single or multiphase), for use in operating AC appliances or supplying power to an electricity grid.

K

kinetic energy   the energy of a body or a system with respect to the motion of the body or of the particles in the system.

L

latent heat   heat energy absorbed or rejected when a substance is changing state and there is no change in temperature.

latent heat gain   the amount of latent heat that is given off by sources inside of a building, which in turn increases the amount of moisture in the air.

lattice tower   a lattice tower is a free-standing framework tower. Lattice towers can be used as electricity pylons, especially for voltages above 100 kilovolts, as radio towers (self-radiating towers or as carriers for aerials), or as observation towers.

lignin   an organic substance that, with cellulose, forms the chief part of woody tissue.

lithium bromide   a type of salt solution used in an absorption chiller.

load calculation   the sum total of BTUs transferred into and out of a building as a result of convective, conductive, and radiant heat transfer.

M

magnetic south   the direction toward which the south-seeking arrow of a compass points.

manifold   a device where multiple outlets or inlets can be controlled with valves or other devices.

modified square wave inverter   a device that converts direct current to alternating current creating a wave form that has a step or dead space between the square waves. This reduces the distortion or harmonics that causes problems with electrical devices.

N

nacelle   the cover for the gearbox, drive train, generator, and other components of a wind turbine.

national renewable energy laboratory (NREL)   a federal laboratory dedicated to the research, development, commercialization, and deployment of renewable energy and energy-efficiency technologies.

net metering   an electricity policy for consumers who own renewable energy facilities through which they can receive credit for the energy they produce.

nickel-cadmium battery   a type of rechargeable battery using nickel-oxide hydroxide and metallic cadmium as electrodes.

nickel-iron battery   a storage battery having a nickel-oxide hydroxide cathode and an iron anode, with an electrolyte of potassium hydroxide. The active materials are held in nickel-plated steel tubes or perforated pockets.

n-type material   a type of extrinsic semiconductor where the doping atoms are capable of providing extra conduction electrons to the host material (e.g., phosphorus in silicon). This creates an excess of negative (n-type) electron charge carriers.

O

ohm's law   states that the current through a conductor between two points is directly proportional to the potential difference or voltage across the two points, and inversely proportional to the resistance between them.

open-loop system   a geothermal system that uses the water in the earth as the heat-transfer medium and then expels the water back to the earth.

P

parallel circuits   an electrical circuit where the voltage across each of the components is the same, and the total current is the sum of the currents through each component.

peak sun hours the equivalent number of hours per day when solar irradiance averages 1 kW/m². For example, 6 peak sun hours means that the energy received during total daylight hours equals the energy that would have been received had the irradiance for 6 hours been 1 kW/m².

pex cross-linked polyethylene pipe. Through one of several processes, links between polyethylene molecules are formed to create bridges (thus the term "cross-linked"). This resulting material is more durable under temperature extremes and chemical attack, and better resists creep deformation, making PEX an excellent material for hot water and other applications.

photovoltaic the process of producing electric current or voltage using electromagnetic radiation from visible light from the sun.

photovoltaic panel a panel used to convert solar energy into electrical energy.

plenum a sealed chamber at the inlet or outlet of an air handler. The duct attaches to the plenum.

pond loop a type of horizontal, closed-loop geothermal heat pump system in which the fluid-filled plastic heat exchanger pipes are coiled and placed at the bottom of a pond.

potable suitable for drinking.

pressure-relief valve a valve designed to open and release vapors or water when a certain pressure is reached.

propylene glycol an antifreeze fluid used in water-type systems.

psi abbreviation for pounds per square inch.

P/T plugs plug-type devices found on water lines that allow the user to take pressure and temperature readings while eliminating the need for leaving gauges or temperature indicators online.

p-type material a type of material that is obtained by carrying out a process of adding a certain type of atoms to the semiconductor in order to increase the number of free charge carriers (in this case positive).

pulse-width-modulation (PWM) controller an electrical controller that shuts off the charging current when the batteries reach a predetermined setpoint by gradually decreasing the charging current pulse as the battery voltage rises.

R

rectifier an electrical device for converting alternating current to direct current, as in a battery charger or converter.

refractory brick a block of refractory ceramic material used in lining furnaces, kilns, fireboxes, and fireplaces. A refractory brick is built primarily to withstand high temperature, but will also usually have a low thermal conductivity for greater energy efficiency.

renewable portfolio standard (RPS) a regulation that requires the increased production of energy from renewable energy sources, such as wind, solar, biomass, and geothermal.

reversing valve a component of a heat pump that reverses the refrigerant's direction of flow, allowing the heat pump to switch from cooling to heating or heating to cooling.

r-value a measure of the capacity of a material to resist heat transfer. The R-value is the reciprocal of the conductivity of a material (U-value). The higher the R-value of a material, the greater its insulating properties.

S

selenium a chemical element with the atomic number 34, represented by the chemical symbol Se. It is a nonmetal, chemically related to sulfur and tellurium, and rarely occurs in its elemental state in nature.

shunt controller a controller that prevents battery overcharge by short-circuiting the PV modules when the batteries are fully charged. The shunt controller circuitry monitors the battery's voltage, and then switches the current flow coming from the solar panels through a power transistor when the fully charged setpoint has been reached.

silicon the eighth most common element in the universe by mass. A substance from which many semiconductors are made.

series circuit an electrical circuit where the current through each of the components is the same, and the voltage across the components is the sum of the voltages across each component.

series-parallel circuit a type of electrical circuit where two or more series circuits are wired together in a parallel configuration.

sine wave inverter an electrical device that converts direct current to alternating current. The converted AC can be at any required voltage and frequency with the use of appropriate transformers, switching, and control circuits.

single-stage series controller a controller used to switch a solar array off when the battery voltage reaches a predetermined value called the charge termination setpoint. This prevents the batteries from becoming overcharged.

slinky loop a type of horizontal, closed-loop geothermal heat pump system in which the fluid-filled plastic heat exchanger pipes are coiled like a slinky to allow more pipe in a shorter trench.

solar altitude angle the angle between the direction of the geometric center of the sun's apparent disk and the (idealized) horizon.

solar gain the amount of energy that a building absorbs due to solar energy striking its exterior and conducting to the interior or passing through windows and being absorbed by materials in the building.

square wave inverter an inverter used to run DC power through switching transistors and then into a transformer.

standing column a vertical recirculating geothermal well. Heat exchange occurs when water is removed from one end of the bore and returned to the other end.

stovepipe the chimney of a biomass stove.

stratification the process where layers of heat rise to the top of a structure.

synchronous generator a type of AC electrical generator that uses the principles of induction motors to produce power. Induction generators operate by mechanically turning their rotor in generator mode, giving negative slip. In most cases, a regular AC asynchronous motor is used as a generator, without any internal modifications.

system grounding the practice of bonding the neutral wiring to earth allowing an alternate path for the electrons to flow thus protecting humans from electrical shock.

T

therm a quantity of heat that is equal to 100,000 BTUs.

thermostatic expansion valve (TXV) a valve used in refrigeration systems to control the superheat in an evaporator by metering the correct refrigerant flow to the evaporator.

thermosyphon system a passive solar hot water system that relies on natural convection to circulate water through the collectors and to the tank. As water in the collector heats, it becomes lighter and rises naturally into the tank above. Meanwhile, cooler water in the tank flows down pipes to the bottom of the collector, causing circulation throughout the system.

torque the twisting force often applied to the starting power of a motor.

transmission loss the heat loss through walls, windows, doors, ceilings, floors, and so forth as a result of heat conduction.

trisodium phosphate (TSP) a colorless water-soluble compound occurring as crystals. It is used chiefly in the manufacture of water-softening agents, detergents, paper, and textiles.

true south the direction, at any point on the earth, that is geographically in the northern hemisphere facing toward the South Pole of the earth. Essentially, it is a line extending from the point on the horizon to the highest point that the sun reaches in the sky on any day (solar noon).

U

un-pressurized systems a solar thermal system where the circulating loop is open to the atmosphere.

utility grid the system of distribution lines that delivers energy from power plants to homes and businesses across the country. The grid is owned and operated by the hundreds of utility companies across the country.

U-value a value that describes the ability of a material to conduct heat or the number of BTUs that flow through 1 square foot of material in 1 hour. It is the reciprocal of the R-value. The lower the number, the greater the heat-transfer-resistance (insulating) characteristics of the material.

V

ventilation loss heat loss as a result of the use of exhaust fans in a building.

vertical axis wind turbine (VAWT) a type of wind turbine in which the axis of rotation is perpendicular to the wind stream and the ground. VAWTs work somewhat like a classical water wheel, in which water arrives at a right angle (perpendicular) to the rotational axis (shaft) of the water wheel.

vertical ground loop a type of closed-loop geothermal heat pump system in which the fluid-filled plastic heat exchanger pipes are laid out in a plane perpendicular to the ground surface.

voltage drop the reduction in voltage of an electrical circuit that is directly proportional to the length of the circuit.

W

water-to-air heat pump a heat pump that uses water as its primary source of heat transfer and air as its secondary source.

water-to-water heat pump a heat pump that uses water as both its primary and secondary source of heat transfer.

wind shear the difference in wind speed and direction over a relatively short distance in the atmosphere.

Y

yaw rotations about the respective axes starting from a defined equilibrium state. The yaw drive on a wind turbine is used to keep the rotor facing into the wind as the wind direction changes.

INDEX